Research for Development

The series Research for Development serves as a vehicle for the presentation and dissemination of complex research and multidisciplinary projects. The published work is dedicated to fostering a high degree of innovation and to the sophisticated demonstration of new techniques or methods. The aim of the Research for Development series is to promote well-balanced sustainable growth. This might take the form of measurable social and economic outcomes, in addition to environmental benefits, or improved efficiency in the use of resources; it might also involve an original mix of intervention schemes. Research for Development focuses on the following topics and disciplines: Urban regeneration and infrastructure, Info-mobility, transport, and logistics, Environment and the land, Cultural heritage and landscape, Energy, Innovation in processes and technologies, Applications of chemistry, materials, and nanotechnologies, Material science and biotechnology solutions, Physics results and related applications and aerospace, Ongoing training and continuing education. Fondazione Politecnico di Milano collaborates as a special co-partner in this series by suggesting themes and evaluating proposals for new volumes. Research for Development addresses researchers, advanced graduate students, and policy and decision-makers around the world in government, industry, and civil society.

More information about this series at http://www.springer.com/series/13084

Alberto Rovetta · Edoardo Rovida

Scientific Knowledge Communication in Museums

Springer

Alberto Rovetta
Politecnico di Milano
Milan
Italy

Edoardo Rovida
Politecnico di Milano
Milan
Italy

ISSN 2198-7300 ISSN 2198-7319 (electronic)
Research for Development
ISBN 978-3-030-09831-5 ISBN 978-3-319-68330-0 (eBook)
https://doi.org/10.1007/978-3-319-68330-0

Printed on acid-free paper

This Springer imprint is published by Springer Nature
The registered company is Springer International Publishing AG
The registered company address is: Gewerbestrasse 11, 6330 Cham, Switzerland

Preface to Part I—Museums as Communication

In the first part of this book, the museum is examined from the point of view of its role and value as a means of communication. The opening two chapters offer some general reflections on museums and examine scientific-technical culture and communication. Such communication can be defined as transmission of scientific-technical contents from a "sender" to a "receiver". This transmission requires definition of the knowledge to be attained by the receiver, starting from the hypothesized initial knowledge. The contents between the starting point and the objective require a careful structuring. Each part of the contents should be transmitted via a "communication unit". In the specific case of museums, such a unit may be referred to as the "museal unit". Subsequent chapters in Part I address more closely the application of the above-mentioned general principles to the museum, by the individuation of the analogies between the communication process and the museum. Specific chapters are devoted to the general configuration of the museum, to placards, to the presentation of objects, and to auxiliary means of communication (e.g., catalog and ticket). The concluding chapter is devoted to an analysis of the role of the museum in scientific-technical education, with the aim of proposing input useful for teachers and students.

Milan, Italy Edoardo Rovida

Preface to Parts II to VII—Dynamic Museums and Social Learning

Often we wonder how a proposal, a project, has been born and developed, why and how it has followed unprecedented paths. The second part of this book (Parts II to VII) focuses especially on the "Dynamic Museums for Social Learning" project, which had its baptism on 2 August 2016, when, in accordance with the Convention of the Ministry of Cultural Heritage and Tourism of Italy and under the patronage of the Politecnico of Milan Foundation, the development test began at five Italian archaeological museums: Altino, Ferrara, Ancona, Egnazia, and Sibari. The immediate response from visitors was unexpected. Within a month more than 1,600 documented visitors had adopted the project.

The project fascinates because it uses computers or tablets or cell phones to explore the mood and views of the visitor, sending this information to the project control center. This center selects from among all the museums that are within the 'gold ring' of the Project, pointing to the pieces that best match the visitor's requirements.

The path that was unfolding in 2016 in fact began long ago. At the end of the Second World War, the nations of Europe, shocked by internal disasters and external crises, were drawn together with the aim of building a caring and rewarding future and had the courage to abandon the old schemes. Some nations started the process of recovery termed the economic miracle while others set about benefiting from the rewards of victory.

Photographs from that era still hit us hard. Children were sometime virtual exiles within a territory. The elderly searched for help, while many young people were ready to move, emigrate, work, and engage. For the "Dynamic Museums for Social Learning" project, a point of reference is part of the critical analysis of a 1971 thesis, listed in the bibliography, which first describes "social learning" from culture, including museums. "Social" is here understood in a modern form, linked to the presence in culture and knowledge (via learning) of all humanity in the same period, in the same world, in the same network. This setting, drawn from a thorough analysis of Anglo-Saxon literature regarding the new frontiers of communication, is the first example of a "social learning" quote and it forms the backbone of the Dynamic Museums for Social Learning project.

The motivation for the birth of the Dynamic Museums for Social Learning project is simple: every museum constitutes a report of individuals' own lives, and like every part of life, it is lived, shared and loved with the individual's own perception and sensitivity. Every visitor must be at the center, the only protagonist of the visit to the museum. The Dynamic Museums for Social Learning project recognizes this fact and keeps the promise to respect it. We are at the beginning of a revolution that is born of social learning. We are about to pass through an era of new artificial intelligence and the outcome will be the sharing, among 7 billion people, of the results of culture. No longer will the museum be a castle isolated from reality, under the influence of unknown spirits.

A video of the presentation of Dynamic Museums project, given on 3 July 2017 in the Senate of the Republic of Italy, at the Commission of Industry, Commerce and Tourism, is now available on YouTube at the address:

https://youtu.be/xCGBVex6BcI

A link also appears within the archive webtv of the Senate of the Republic of Italy:

http://webtv.senato.it/4622

Link to the Platform of DYNAMIC MUSEUMS

http://www.museidinamici.cloud

Milan, Italy Alberto Rovetta

Acknowledgements

Author Alberto Rovetta would like most of all to thank Anna, his wife, Chiara, his daughter, and Francesco, his son, for their continuous support in the face of the new steps in artificial intelligence for the development of culture and science, devoted to the creation of an increasingly happy way of daily life.

He would also like to thank Prof. Edoardo Rovida and Dott. Giulio Zafferri for their great contribution in revealing all aspects of the new museums.

Many thanks go to the Fondazione Politecnico di Milano and to the Politecnico di Milano, Department of Mechanics, and Faculty of Design, for their cooperation.

Deep thanks go to Sen. Massimo Mucchetti for his scientific definition of progress and to Dr. Ugo Soragni, Arch. Manuel Roberto Guido, and Dott. Valentino Nizzo, of MiBACT, for their support in convention work.

Grateful thanks for their cooperation goes to the Directors of the Archaeological Museums of Altino, Ferrara, Ancona, Egnazia, Sibari, of MiBACT, as well as for the passionate and strong contributions of Dr. Domenico Piraina and Dr. Simone Percacciolo, Comune di Milano, and, from the Museo Marinaro di Camogli, Com.te Bruno Sacella, and Mr. Mario De Bernardi.

Grateful thanks go to Mr. Roberto Biscuola, Servizio Web e Grafici di Ateneo, Politecnico di Milano, for the great and skillful cooperation in graphic improvement.

Thanks also go to Mr. Luigi Candita, for his strong cooperation in designing the app.

The author would also like to tip his hat to the work of Prof. Emilio Massa, Past Rector of Politecnico di Milano, who first introduced congresses on neurology at the Politecnico, opening our eyes to the incredible results for a new world of communication to be found through artificial intelligence.

Alberto Rovetta

Author Edoardo Rovida would like very much to thank his wife Rita for her support. Many thanks also go to his colleagues and friends, Prof. Alberto Rovetta, and Dr. Giulio Zafferri.

The author is also very grateful to the Politecnico di Milano, to the Department of Mechanics (and particularly to the Director, Prof. Marco Bocciolone), to the Fondazione del Politecnico (Prof. Giampio Bracchi and Engineer Daniele Bignami) and to MiBACT (Ministero dei Beni ambientali culturali e del turismo) (Dr. Ugo Soragni, Arch. Manuel Roberto Guido, and Dr. Valentino Nizzo).

Thanks also go to the MULSA (Lombard Museum of History of Agriculture) (Prof. Luigi Mariani) in S. Angelo Lodigiano, Lombardy, to the Leonardo 3. Il mondo di Leonardo in Milan (Engineer Massimiliano Lisa and Dr. Edoardo Zanon), to the Clup (publishing house of the catalog of the exhibition for the 125th anniversary of the Politecnico di Milano), and to the La Goliardica pavese (publishing house of the monograph "La fisica a Pavia nelle opere di Giuseppe Belli").

Particular thanks go to Dr. Marina Carulli for his careful collaboration and to Dr. Sara Calabrò and Dr. Giorgio Menzio for their long-term collaboration in the field of cultural heritage.

Edoardo Rovida

Contents

Courtesy of Ferrara Archaeological Museum

Courtesy of Ferrara Archaeological Museum

Courtesy of Sibaritide Archaeological Museum

Part I
Museums as Communication

Chapter 1
Scientific-Technical Culture and Museums

1.1 The Culture

Culture is a word whose etymology springs from the Latin word *colere*, which means grow; that is, it suggests implementation of the human being. Today, the word culture means a set of knowledge, opinions, ability, and/or behavior that characterizes and upgrades the human being. The original etymology, therefore, is in perfect accordance with the modern sense of the word.

Culture, in general, is relative to each object and each aspect of human activity and behavior within it. In the context of the present text, of course, culture is relative to each scientific-technical subject, i.e., scientific principles and relative applications for realizing machines, instruments, and devices with the aim of performing functions useful for human beings.

Scientific-technical culture plays an important role in modern life: in effect, each person is a user of technical products, and therefore, it is fundamental that we cultivate the ability to utilize such products safely.

In a general sense, it is possible to observe that scientific-technical culture, beyond the notion of culture in a wider sense, is characterized by four components:

1. the informative component, related to the acquisition of knowledge. Such a component is, of course, related to the transmission of said knowledge, which must reach the user in some way;
2. the critical component, with the aim of developing a critical understanding of the information obtained in the previous component;
3. the practical component, which involves the development of the ability to apply the acquired information to concrete situations and the solution of problems;
4. the deontological component, related to the correct and ethical utilization of the above acquired skills.

A. Rovetta and E. Rovida, *Scientific Knowledge Communication in Museums*,
Research for Development, https://doi.org/10.1007/978-3-319-68330-0_1

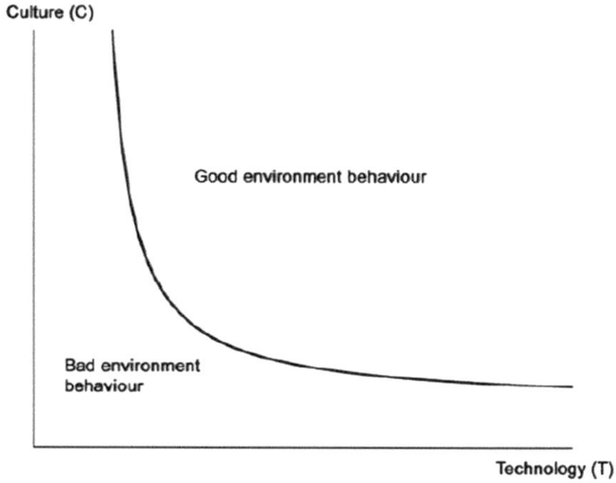

Fig. 1.1 Relation between culture and technology in terms of the environment

In other words, culture means a mix of skills (informative, critical and practical components) and deontology (the ability to address human behavior in correct relation to a given ethical aspect).

The important role of culture is depicted expressively in Fig. 1.1 (Rovida and Viganò 2006). In this figure, the influence of culture and technology is related to environmental aspects. It is easy to recognize that the same diagram could also be valid, e.g., in regard to safety.

To cultivate good behavior in regard to the environment (or, respectively, safety), it is necessary to improve the technology and/or culture. It is important to observe that an upgrade in culture could compensate for a decline in technology. Of course, the most important result is achieved when culture and technology are upgraded contemporaneously. The scientific-technical culture of today is often, unfortunately, considered to be less important in comparison with the humanistic culture. In fact, it represents a single culture, with two “faces”: the humanistic one and the scientific one. The scientific “face” needs to be implemented, but, of course, without forgetting the humanistic implications.

As will be observed later, the role of scientific-technical museums in the development of knowledge, deontology and technology is fundamental.

1.2 Role of Scientific-Technical Heritage in the Maintenance of Scientific-Technical Culture

Scientific-technical heritage can be defined as the set of tangible and intangible artifacts that represents the state of the art of the scientific-technical development in a specific period of time and in a specific field. Today, our scientific-technical heritage can be defined as the sum total of scientific-technical knowledge gathered since the origin of our ability to understand such things until the present time.

Examples of such artifacts can be machines, instruments, devices, technical products, industrial and civil buildings, factories, models, software, books, documents, technical drawings, and so on.

Our scientific-technical heritage represents a witness to the history of the human mind, to the development of scientific-technical knowledge, and to the respective revelations from each period of time. This scientific-technical heritage plays an important role from many points of view. Here, we examine some of the most important of these.

Historical revelations are key to reading the history of science and technology, which is an important part of the overall History.

Figure 1.2 is a picture related to agricultural life in the Middle Ages (MULSA, Lombard Museum of History of Agriculture, S. Angelo Lodigiano, Lombardy). Such a picture, with its representation of certain objects and actions, can therefore be regarded as a scientific-technical document, because, in effect, it represents a specific moment of life in a food shop in the Middle Ages.

It is interesting to observe that in many artworks, technical objects are often present. In such artworks, the represented technical objects bear witness to the technical state-of-the-art of the time. Therefore, from this point of view, these artistic representations are important: be they ancient paintings, mosaics, stained glass, etc., when they represent machines or other technical objects, they recount historical constructive solutions, technical activity, and working organization. Scientific-technical heritage, therefore, can also be documented through artistic representations.

Fig. 1.2 Example of agricultural life in the Middle Ages (courtesy of MULSA, Lombard Museum of History of Agriculture, S. Angelo Lodigiano, Lombardy)

Another example is the reconstruction of the Tower of Filarete in Castello Sforzesco in Milan (destroyed by an explosion in the Sixteenth Century): the prominent Italian architect Luca Beltrami (principally active in the Twentieth Century) was inspired in the reconstruction by some graffiti in the Abbey of Clairvaux, near Milan.

In the Museo del Duomo (Museum of the Cathedral Treasure) at Vigevano (a town in Lombardy, near Milan), there is a tapestry in which some unknown musical instruments are represented. Such representations, therefore, could be a basis for research on the history of music.

Another consideration is that historical revelations are often the basis for innovations: e.g., old ideas abandoned as a result of material and technological problems can be "reinvented" today under modern conditions. For example, the stern rudder was "invented" by the Egyptians, and then "rediscovered" in the Thirteenth Century.

The Roman politician Cicero (106–43 B.C.) said, "*To be ignorant of what occurred when you were born is to remain always a child.*"

Leon Battista Alberti (1404–1472), another prominent Italian architect, wrote "*They have preserved examples of ancient works, such as theaters and temples, from which, as from great teachers, much can be learned.*"

One significant example is that of the Conservatoire (the first modern scientific-technical museum, founded in Paris in 1794), which had been proposed by Descartes (Seventeenth Century) with the idea of presenting the evolution of machines in chronological order, to stimulate the creativity of designers and technicians. This example stands as early confirmation that critical consideration of historical constructive solutions can play an important role in the development of innovation.

A Portuguese proverb asserts "*A country, like a plant, becomes great if it can protect and preserve its roots.*"

Figure 1.3 represents an electrical device created by Giuseppe Belli, an Italian Physicist (Nineteenth Century) and professor at Pavia University: the drawing anticipates the use of arrows relative to the mutual positions of the orthographic projections, which has become so commonplace in modern times (Bellodi and Rovida 1994).

Figure 1.4 represents a German patent (1943) that anticipates hydro-pneumatic suspension, successively adopted in the 1950s. Such a constructive solution allowed for the vertical levelling of the car body (Biggioggero et al. 2003).

Aesthetics are also of interest in many historical constructive solutions. As an example, Fig. 1.5 represents two drawing instruments (Nineteenth Century) that are characterized by great beauty.

Design solutions of the past can also stand as examples of professionalism, skills, ethics, etc. The drawing instruments in Fig. 1.5 are examples of beauty and professionalism.

The consideration of historical scientific-technical heritage can also be interesting from the educational point of view. The telling of the story of how any given historical revelation came about can result in an appreciation of the process of

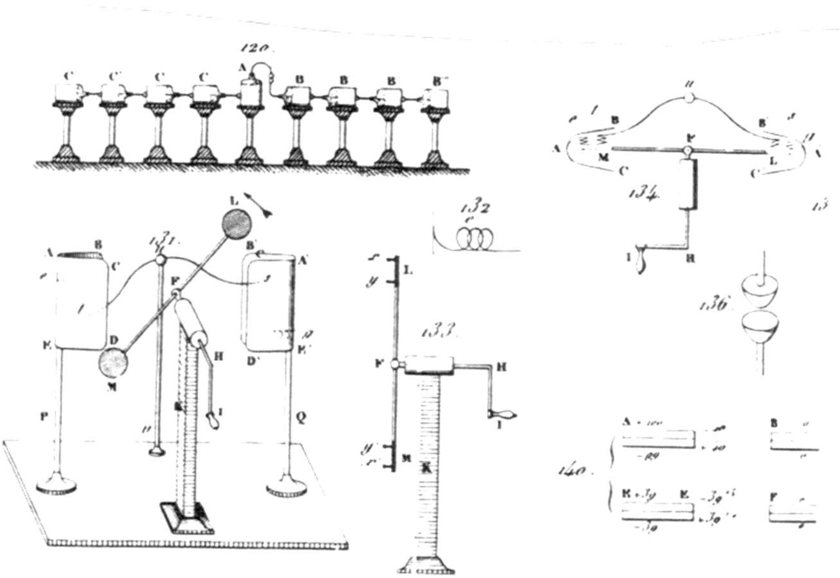

Fig. 1.3 Drawing by Giuseppe Belli (Nineteenth Century): an electrical device that anticipates the "arrow method" of orthographic projections

Fig. 1.4 Historical (1943) hydropneumatic suspension that anticipates modern constructive solutions

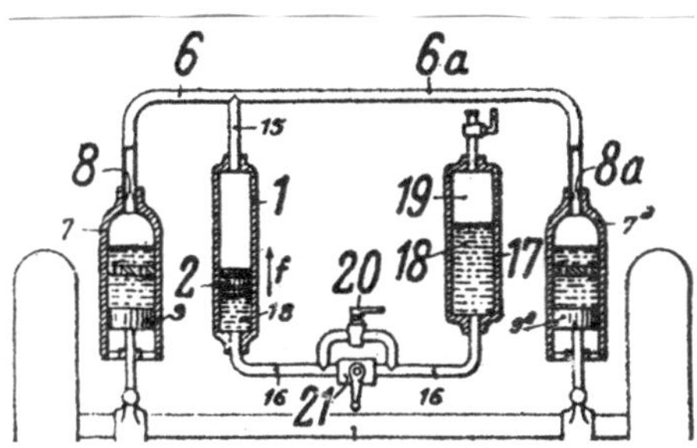

positive development, an assessment of the work that lies beneath every achievement, which may seem "trivial" today, and correct evaluation of current developments, which may, of course, be superseded by new ones in short order. In addition, historical constructive solutions very often turn out to be simple and easily individuated, allowing basic principles to be explained to students.

Last, but not least, historical heritage is interesting from an economic point of view, even in the scientific-technical field: museums, e.g., are often businesses that generate large amounts of capital. It is sufficient to think of the upgrade in tourism that can be brought about by the presence of museums, as well as the resultant upgrade in jobs.

The world's historical scientific-technical heritage (and particularly that of Italy) is gigantic. The tools for scientific-technical historical culture are, in general, as

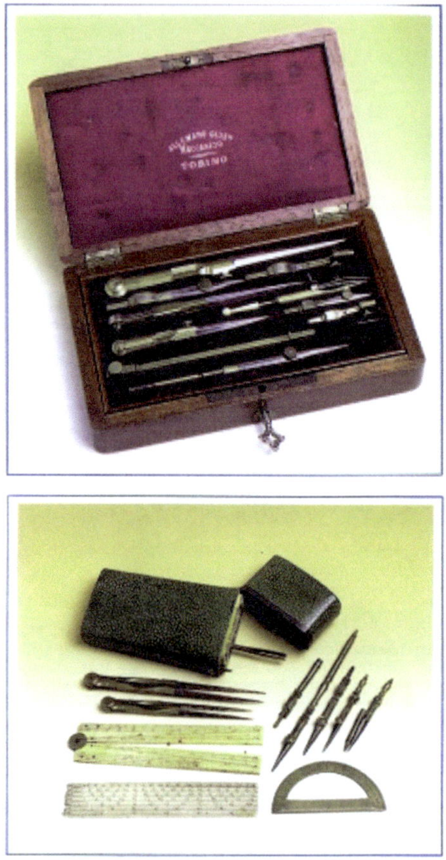

Fig. 1.5 Drawing instruments (Nineteenth Century), characterized by beauty

follows: libraries, archives, museums and private collections. In this text, we focus on the museums.

1.3 Museums

The official definition of a museum has been given by the ICOM (International Council of Museums), an international association founded in 1946 that brings together curators, museologists and cultural operators (Sutera 2016).

According to the ICOM Statutes, adopted during the 21st General Conference in Vienna, Austria, in 2007:

> A museum is a non-profit, permanent institution in the service of society and its development, open to the public, which acquires, conserves, researches, communicates and exhibits the tangible and intangible heritage of humanity and its environment for the purposes of education, study and enjoyment.

This definition stands as a reference throughout the international community.

Another definition, unofficial, but significant, is "*the museum is the place where time becomes space.*"

Among the many types of museum, the following can be considered the most fundamental:

- artistic
- archeological
- scientific-technical
- scientific-natural history
- ethnographic
- historical.

The expression *scientific-technical museum* can be extended to include, for example, ethnographic museums. In other words, while the scientific-technical museums addressed in the present text are generally those museums devoted to scientific-technical heritage, it is necessary to reconsider a concept mentioned above: each field of knowledge has scientific-technical components. An art museum can have pictures that represent machines; such pictures are scientific-technical documents. In addition, an art museum, for example, very often utilizes scientific-technical means to better observe the exposed material, such as the digitalization of pictures, virtual reality, and multimedia applications.

Scientific-technical museums can also be classified into certain subcategories. The following can be recognized as being among the most important and interesting:

1. general, i.e., museums devoted to many scientific-technical fields, such as the Museo Nazionale della Scienza e della Tecnologia (National Science and Technology Museum) "Leonardo da Vinci" in Milan, the Deutsches Museum (German Museum) in Munich, and the Science Museum in London;
2. specialized museums, i.e., museums devoted to specific aspects or objects of science and technology, such as the Museo dell'Automobile (Automobile Museum) in Turin, the London Transport Museum, the Swiss Transport Museum (*Verkehrshaus der Schweiz*) in Lucerne, the Museo Lombardo di Storia dell'Agricoltura (Lombard Museum for the History of Agriculture) in S. Angelo Lodigiano, near Milan, and the Museo della Tecnica Elettrica (Museum of Electrical Technology) in Pavia;
3. integrated museum systems, i.e., an organic complex of specialized museums. Examples of such museums could be an aircraft museum in an airport, a railway museum in a station, an automobile museum in a disused car factory, etc.;
4. museums for children, i.e., an entire museum (or part of a general museum) with the specific aim of teaching such subjects as physics or the chemical sciences to children through educational forms of play, e.g., some of the specific laboratories in the Museo Nazionale della Scienza e della Tecnologia (National Science and Technology Museum) "Leonardo da Vinci" in Milan;

5. university museums, i.e., the collections donated to the universities founded in Italy in the Eighteenth Century, largely utilized as tools for scientific and didactic activity, and which today stand as a valid document of the history of the university, e.g., The University Museum in Pavia and the historical museum of Politecnico di Torino (Rossetti and Sassi Perino 1989);
6. company museums, i.e., collections of the products realized by an industry. Such museums, in many cases, are considered to be specialized museums.
7. distributed museums, i.e., a complex integration of museums and landscape in a specific territory;
8. open-air museums, i.e., complexes of houses and constructions (original or reproduced) that represent, for example, villages, with the aim of documenting the life of a country in a certain period of time. One example of such a museum is the Skansen Village in Sweden, founded in 1891.

In addition to the museums that are defined, as specified above, as institutions devoted to conservation and exposition, there are other possibilities for cultural heritage, related in particular to the idea of the distributed museum. For example, towns, in many cases, are essentially collections of interesting historical objects, documenting such things as forms of architecture and modes of construction. Entire sections of a town may be characterized by specific historical interests and bear witness to the abilities of our ancestors and their technical knowledge.

Cooperation between museums is also very important. Two concepts in particular should be highlighted (Cataldo 2016):

- the museum system. This is a top-down organizational structure for the purpose of resource optimization. These can be divided into the following subcategories:

  (a.1) Territorial management systems: these are born already formed and represent coordination of already existing structures.
  (a.2) Integrated cultural territorial: these tend to come about from an already existing reality that converges into a hierarchical structure, based on a central museum and some peripheral museums.
  (a.3) Territorial designed cultural: there are no existing structures and the goal is the design of a system from scratch.

- museum network. These are peer structures for predominantly cultural purposes. The network can be articulated at different levels: scientific, popular, public reception.

1.4 Some Considerations About History

The following considerations are particularly relative to scientific-technical museums.

The first origins of the museum can be found in the tendency of the human being in prehistoric times to collect objects of different types. The word 'museum,' from the old Greek *museion*, means *house of the Muses*, where the Muses protect the arts and sciences, i.e., collections of artworks and tools, in order to preserve the tradition. The ancient architect Vitruvio (First Century B.C.) recognized that, in addition to spiritual values, there were technical and scientific values of equal dignity.

Tolomeo II Filodelfo (Third Century B.C.) founded the Mouseion (set of libraries and collections of instruments available to scientists) in Alexandria. A particularly interesting part of the Mouseion is the Library of Alexandria, the biggest collection of books of the ancient world. Linked to the Library itself, there were rooms for study, laboratories and astronomical observatories. Many scientists of antiquity studied in the Mouseion, including Euclid, Archimedes, Eratosthenes, Hipparchus, Pappus, Hypatia and Aristarchus.

The Renaissance saw the rise of the so-called *Wunderkammer* (in German, *rooms of wonders;* also called Kunstkabinett, Kunstkammer, Cabinets of Wonder, and Wonder-rooms), collections owned by princes and scientists that included *naturalia* (animals, minerals, fossils) and *artificialia* (machines, tools). Particularly important examples of Wunderkammer were those owned by von Rudolf of Hapsburg (1552–1612), Maria Luisa de' Medici (1667–1743) and Manfredo Settala (son of Ludovico, the physician of the plague described by Alessandro Manzoni). In 1751, the latter collection, located in Milan, was included in the Ambrosiana Library (Biblioteca Ambrosiana); it was subsequently destroyed by the bombings that devastated that city in 1943. The only remaining pieces of the Wunderkammer of Manfredo Settala is an automaton conserved in the Natural History Museum (Museo di Storia Naturale) in Milan and some armillary spheres, conserved in the Ambrosiana Library (Biblioteca Ambrosiana).

The first scientific-technical museums in the modern sense began to come about in the Seventeenth and, particularly, Eighteenth Centuries. The most important example is the Conservatoire opened in Paris in 1794. It was proposed by Abbé Henry Grégoire (1750–1831) as a "depository for machines, models, tools, drawings, descriptions and books in all the areas of the arts and trades". As mentioned earlier, the French philosopher Descartes first proposed the idea of presenting the evolution of machines in chronological order for the purpose of stimulating creativity. Thus, it is reasonable to observe that Descartes, through contemporary means, proposed the very first "database."

In the Nineteenth Century, some great scientific and technical museums were born (often under the auspice of Universal Expositions): the Science Museum of London in 1851, the Technisches Museum in Vienna in 1873, etc.

In 1906, an Exhibition was held in Milan celebrating the Simplon Tunnel. At the closure of the event, Giuseppe Belluzzo (a teacher at the Politecnico di Milano) asserted that, in Italy too, the constitution of a museum documenting the stages of science and technology would be an important achievement.

Other examples include:

- 1903: foundation of the Deutsches Museum in Munich.
- 1931: Guglielmo Marconi, President of Consiglio Nazionale delle Ricerche (National Council for Researches), takes up the idea of the Belluzzo.
- 1942: Guido Ucelli, manager of Riva Calzoni, establishes the Foundation for the National Museum of Science and Technology, the first step in the constitution of the National Museum of Science and Technology. In 1953, the National Museum of Science and Technology opened in Milan with a big Exhibition about Leonardo da Vinci, for the 500th anniversary of the birth of Leonardo.

The 1970s saw the birth of many new museums, especially specialized ones. In addition, many demo-anthropological museums, devoted to agriculture, arts, and crafts, were begun.

Another characteristic of this time was an upgrade of the relationship between the exposed materials and the visitor so as to make it more active, no longer limited to simply observing, but actually doing something in regard to the exposed material. This interaction was, initially, only mechanical; in other words, the visitor would act, e.g., pull a lever or a pulley, so as to be able to examine the functioning of a mechanism. Successively, the interaction became electronic, and then computerized.

Nowadays, the impact of virtual and augmented reality in museums is very strong and rapidly increasing.

Another trend is the introduction of the Science Center, characterized by an introduction to science characterized by entertainment values, sometimes excessively spectacular, which, in the 1980s and '90s, began to be assisted by the development (also sometimes excessive) of multimedia. Another trend is the technology park, concentrated in limited areas within universities, museums, and research centers.

An additional trend has been the creation of an integrated system of specialized and coordinated museums, located in specific areas: for example, a railway museum in a station, an aeronautical museum in an airport, if possible, connected by a transport line.

Modern times have also seen the creation of the eco-museum and the open-air museum (installations that reconstruct or safeguard an environment or a portion of territory).

The ICOM, created in 1946, is the only organization of museums and museum professionals with a global scope, committed to the promotion and protection of natural and cultural heritage, present and future, tangible and intangible. With approximately 30,000 members in 137 countries, the ICOM represents a network of museum professionals acting in a wide range of museum-and heritage-related disciplines.

Table 1.1 presents a classification of the periods of evolution of museums.

Table 1.1 Classification of periods of museums

Period of time	Specific events
Prehistoric and ancient times: the origins	Natural trend of the human being towards collecting interesting and curious objects
Fifteenth Century: Humanism	Collection of objects by princes and scientists
Sixteenth and Seventeenth Centuries: the Wunderkammer	The collections of princes and scientists become more available
Eighteenth Century: the first modern museums	The Conservatoire in Paris is born
Nineteenth Century: the great European museums and exhibitions	Foundation of the great modern science museums, sometimes in the wake of the closure of great international exhibitions
Twentieth Century (1900–1950)	Upgrade in the number of scientific-technical museums
Twenty-first Century	Upgrade in the interaction between visitors and the materials exhibited Upgrade in the application of multimedia and virtual reality

1.5 Disciplines

Some disciplines are related to the design, realization and management of museums: museology, museography and museumtechnique. Such disciplines are characterized by many definitions, not always uniform. In particular, the words *museology* and *museography* are often used indiscriminately.

Among the many definitions, the proposal of the Authors, as will be represented in the present text, is as follows:

1. museology: formalization of the general ideas for the basis of the museum and transformation of general ideas in the design of the museum
2. museography: transformation of the design of the museum in the product (i.e., the museum)
3. museotechnique: management of the museum.

1.6 Some Considerations About Ethics and Cultural Heritage

With an increase in institutions, the requirement for rules and behavioral principles always becomes more important. Such rules are particularly devoted to the correct management of the cultural heritage. In aa.vv. (2016), the experiences of many European Countries are presented and compared.

Ethical principles also play a fundamental role in the field of cultural heritage. From such a point of view, the activity of the ICOM is very important. The eight fundamental principles of the ICOM Code are presented here.

1. Museums preserve, interpret and promote the natural and cultural inheritance of humanity.
2. Museums that maintain collections hold them in trust for the benefit of society and its development.
3. Museums hold the primary evidence for establishing and furthering knowledge.
4. Museums provide opportunities for the appreciation, understanding and management of natural and cultural heritage.
5. Museums hold resources that provide opportunities for other public services and benefits.
6. Museums work in close collaboration with the communities from which their collections originate, as well as serving those communities.
7. Museums operate in a legal manner.
8. Museums operate in a professional manner.

References

aa.vv. (2016) In Murphy BL (ed) Museums, ethics and cultural heritage. Routledge Taylor & Francis Group

Bellodi G, Rovida E (1994) La fisica a Pavia nelle opere di Giuseppe Belli. La Goliardica Pavese, Pavia

Biggioggero GF, Calabrò S, Menzio G, Rovida E (2003) Evoluzione storica di soluzioni costruttive. Il caso di sospensioni per autovettura Politecnico di Milano—Dipartimento di Meccanica CNR (National Council of Researches) Progetto finalizzato Beni culturali

Cataldo L (2016) Musei e patrimonio in rete. Hoepli, Milano

Rossetti U, Sassi Perino AM (1989) The role of the historical museum of the Politecnico di Torino in education and in the conservation of the local cultural heritage. Paper presented at SEFI (European Society of Engineering Education) annual conference, Naples, 17–20 Sept 1989

Rovida E, Viganò R (2006) Communication for X(CFX): a contribution to the design for sustainability. In: Proceedings of the 7th international conference on technology and quality for sustained development, Bucharest, 25–27 May 2006

Sutera S (2016) ICOM Italia ICOM (International Council of Museums) 2016

Chapter 2
General Principles of Scientific-Technical Communication

2.1 Logical Schema of Communication

The museum, and particularly, although not exclusively, the scientific-technical museum, is a form of scientific-technical communication. Many considerations about communication, therefore, can be easily applied to museums.

It is useful to declare that, in the present consideration, the term "communication" means "scientific-technical communication."

Communication can be considered as the difference between two states, an initial state and a terminal state, that represent the amount of knowledge held, respectively, before the communication and after the communication. Because the human being behaves in relation to the knowledge, it is easy to recognize that the initial knowledge can be called "initial behavior" and the terminal knowledge, "terminal behavior." The initial behavior, therefore, is identified according to the prerequisites and the terminal behavior according to the objectives of the communication (Fig. 2.1).

Complete determination of the initial and terminal behaviors is fundamental to designing and realizing optimal communication, and therefore, it is also fundamental to museums.

The aim of communication is to allow us to effect the difference between the terminal and initial behaviors. This difference represents the content to be transmitted. Such content can be transmitted by association with a system of signs (corresponding, e.g., to oral, written, or graphical communication). The perception of the system of signs by the user of the communication represents the received content, in other words, the upgrading of behavior (Fig. 2.2).

The system of signs, in the case of a museum, is very complex, because it is an integration of written, oral, graphical, and expositive communications, which requires very accurate design and realization, with a strong connection between the neurological and psychological aspects.

A. Rovetta and E. Rovida, *Scientific Knowledge Communication in Museums*, Research for Development, https://doi.org/10.1007/978-3-319-68330-0_2

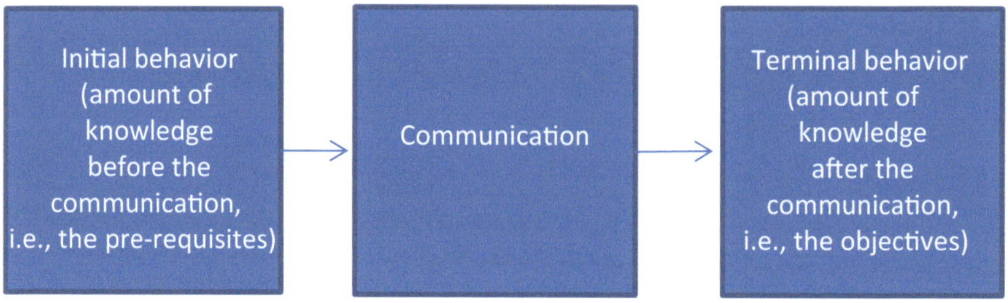

Fig. 2.1 Communication as step between a starting point (initial behavior) and arrival point (terminal behavior)

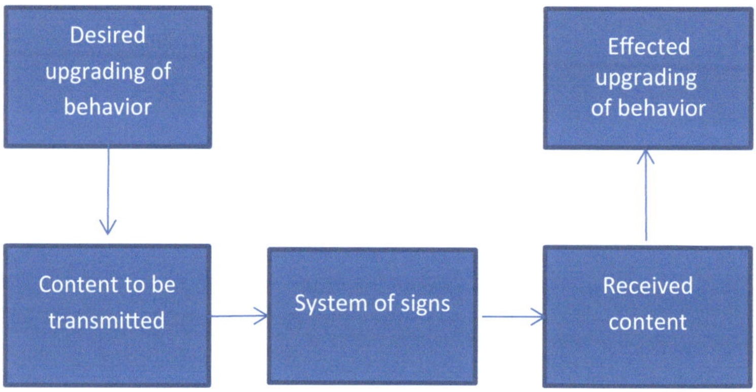

Fig. 2.2 Simplified logical schema of the communication

2.2 Prerequisites and Objectives

Prerequisites and objectives, as mentioned above, are the starting and arrival points of each communication and, therefore, of the museum. Before communicating, it is necessary to know the people to whom the communication is directed and what knowledge they must acquire through the communication. Having taken these steps, it is possible to determine what needs to be communicated (i.e., the content of the communication) and the best way to communicate it (i.e., the style of the communication).

The terminal behavior, at least in the scientific-technical field, in accordance with psychological and pedagogical theories, can be thought of as constituting three fundamental components: the informative, the critical and the practical. It is easy to recognize that the informative component has the aim of providing the user with certain valuable information, while the critical component is devoted to constituting a guide for critically analyzing the information received in the previous phase. Finally, the practical component allows the user to develop the ability to apply the information received and critically analyzed in the first and second components to

Table 2.1 The components of behavior

Component	Aim	Name of the component	Abbreviation of the name of the component
Informative	Passing of information	Rule	RUL
Critical	Encouragement of critical analysis of the received information	Complete example	EG
Practical	Application of the received information in a practical situation	Incomplete example	EG′

practical situations. In addition, it is possible to recognize a deontological component, which aims to steer the effected behavior towards ethical utilization. In the design and realization of a scientific-technical museum, it would be useful to take into account the above-mentioned components (Table 2.1).

2.3 Knowledge to Be Transmitted

The amount of knowledge between the initial and terminal behaviors represents the conceptual complex to be transmitted. Such a complex can be considered to be constituted by elementary portions called *concepts*, which can be defined as blocks of knowledge characterized by a name and a definition. The concepts can be highlighted through a process of analysis of the terminal behavior. Characteristics of sets can be applied to the concepts. Figure 2.3 represents an example of analysis of the terminal behavior (TB): such a process allows us to highlight the component concepts A, B and C. Concept A can be considered to be constituted by the component concepts A1 and A2, while A2, through a successive process of

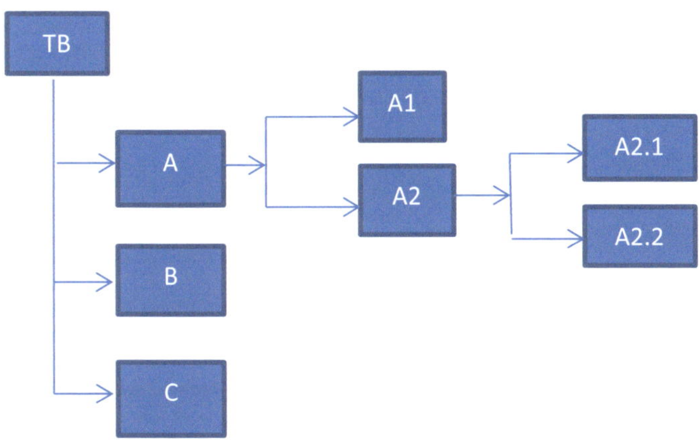

Fig. 2.3 Example of the process of analysis of the terminal behavior (TB), to highlight the components concept

analysis, highlights the component concepts A2.1 and A2.2. The concept A is, of course, the most general: the concepts A1, A2.1, and so on, are more and more particular. It is important to verify, taking into account the sets of characteristics, that the sum of A, B and C corresponds to the TB, and that of A1 and A2 corresponds to A, and, similarly, the sum of A2.1 and A2.2 corresponds to A2.

By following the most modern psychological theories, it is important that the museal unit influence, in particular, the emotional brain hemisphere, e.g., by presenting, before the rule, an "emotional" example capable of upgrading the emotion, the curiosity, the fear of the visitor, etc.; this aspect is particularly important in the complete and incomplete examples.

For example, a scientific-technical museum could present an interesting and unusual machine, while a natural history museum might present the skeleton of a prehistoric animal.

For information on the basics of communication, see Shannon and Weaver (1962), Biggioggero and Rovida (1977) and Biggioggero et al. (2003).

2.4 Museal Unit

The museal unit can be defined as the elementary part of the museum that aims to transmit each concept. Such a unit must facilitate the visitors' effecting of the "terminal behavior" relative to the concept. Therefore, the museal unit should be

Table 2.2 Examples of the configuration of a museum unit

Parts of the terminal behavior	Elements of the parts
RUL	Written text in a poster Written text on a display Oral description presented by a person Oral description presented by a device
EG + EG′	Fixed image Image in motion Fixed real object Real object in motion Real interactive object Fixed virtual object Virtual object in motion Virtual interactive object Real fixed work environment Real work environment in motion Virtual fixed work environment Virtual work environment in motion Virtual interactive work environment Audio Demonstration of utilization of the object And so on

constituted of the components that constitute the terminal behavior, i.e., the RUL (informative component) and the EG and EG′ (critical/informative component). Each of the above-mentioned components can be constituted of many elements. Table 2.2 presents some examples of such elements.

The above-mentioned elements can be presented and combined so as to achieve the following aspects of a museum unit:

- Real object
- Environment
- Landscape
- Virtual models
- Placard
- Historical documents
- Video with historical documents and/or upgraded information
- Three dimensional reconstructions (real or virtual) of people related to the arguments, e.g., scientists, technicians
- Related documents, such as drawings, photos, bills, stamps, cards, flyers, advertising documents, notes by writers and poets, paintings
- Artistic objects related to the concept, e.g., a painting representing a machine or a work experience
- Recorded materials, e.g., songs of birds in a natural history museum, mechanical workshop sounds in a technical museum, voices of scientists providing explications
- "Living museums" (very important and significant from the educational point of view), i.e., the utilization of machines conserved in the museum for the demonstration of their practical use (e.g., racing with historical cars, agricultural operation with historical agricultural devices)
- Original drawings
- 3D drawings in significant dimensions
- Real models
- Videos with assembly and disassembly
- Videos with the possibility of inner observation.

Of course, all of the above-mentioned parts should be accompanied by explications.

2.5 Museum Paths

A museum's visitors can be characterized according to their difference in interest in relation to the exhibited materials: mere curiosity, normal interest, exceptional interest. In relation to the above-mentioned differences in interest, it is necessary to have a corresponding variety of paths through the museum that the visitor can follow during the visit.

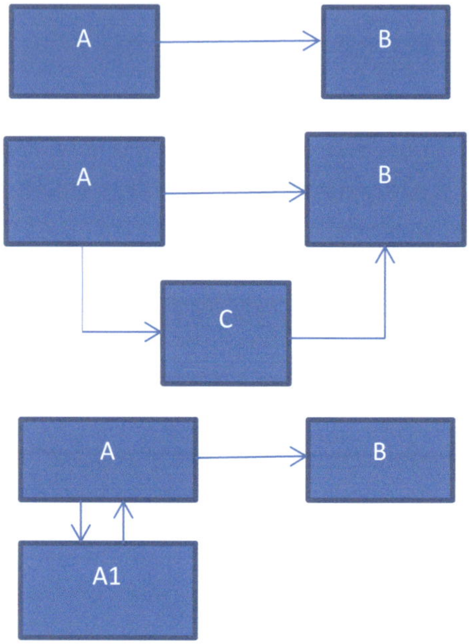

Fig. 2.4 Examples of different paths

Figure 2.4 depicts some examples of a museum's paths. It is easy to recognize the linear path (A, B) for the visitors with only simple curiosity, and the deeper paths (C, A1) with a greater amount of information for more interested visitors. For very interested visitors, it would be possible to propose other "paths," such as a visit to the libraries or to the stores of the museum or contact with the scientific staff.

The linear path and the path featuring a deeper degree of information could be realized in different ways, such as:

- large placards with both general descriptions and individual parts highlighted for the specialized explication of the exhibited objects, e.g., with different colors or other different graphical configurations;
- different showcases in individual rooms;
- different paths through the museum.

Let us look a little more closely at the first idea. At the beginning of each path, a large placard can be placed with a general explanation about the path.

As an example, a part of path can be proposed as in Fig. 2.5.

The different paths can be indicated physically, but also in an ideal sense. As an example, for the 150th anniversary of the founding of the Politecnico di Milano, an exhibition was organized at the Museo Nazionale della Scienza e della Tecnologia (National Museum of Science and Technology) in Milan, with the aim of highlighting the contributions by the teachers and (old) students of the Politecnico to the field of engineering. To that end, some specialized paths were organized, featuring machines, instruments and devices realized through the ideas and contributions of

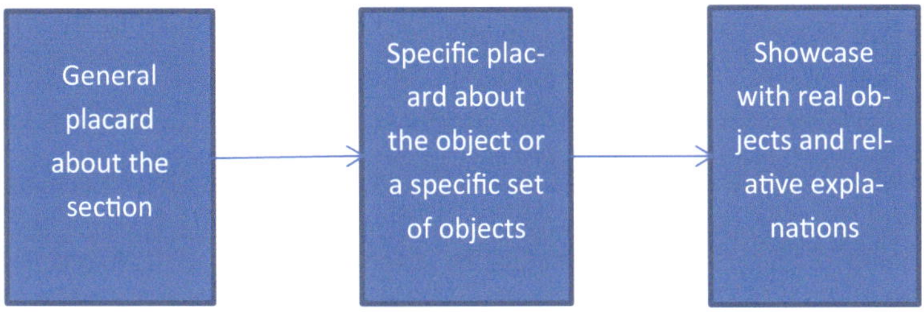

Fig. 2.5 Example of part of a path

teachers and graduates of the Politecnico. This particular emphasis was realized through the use of placards and captions near the above-mentioned machines, instruments and devices.

Paths can also be delineated in relation, e.g., to the interest of the visitor or to the amount of disposable time. For example, if the visitor only has two hours to spend on the visit, through the use of flyers or computer displays, the best route for seeing the most interesting objects in the given time can be suggested.

Similar means can be used to suggest routes/plans in regard to specific categories of visitors, e.g., children.

The paths do not have to confine themselves to only one museum, but can stretch through several museums, identifying the most significant pieces in each of them. The paths can also be at a higher level, stretching between the monuments or museums of a city, or even through different cities.

2.6 Case Studies Relative to the Museum

2.6.1 The Name

The name is the first thing about the museum that everyone hears: therefore, determination of the name plays an important role in the general orientation of the visitors to the space. In general, it is possible to consider the following requirements for the name:

- Immediate impact
- Conciseness
- Image validity
- Expression of the contents of the museum
- If possible, facility to be read in foreign languages or to be translated.

There are, in general, two different approaches to the construct of the name:

- title and subtitle (the first one providing an immediate impact, with the second one explaining the contents)
- title alone.

As an example, in 1999, one of the Authors collaborated in the design and organization of an exhibition celebrating the 200th anniversary of the invention of the battery by Alessandro Volta. The main concepts that needed to be expressed by the title (and eventually by the subtitle) were *battery* and *200th anniversary* (aa.vv 1999).

Some proposals for title were as follows:

- 200 years of the battery
- Battery: the first 200 years
- Celebration of 200 years of the battery
- 200 years of sparks
- Alessandro Volta and the battery: 200 years of sparks
- 200 years ago, the battery
- The battery by Volta: a two-century-long spark.

The definitive choice was as follows:

Title: *a two-century-long spark*
Subtitle: a *commemorative exhibition of 200 years of Volta's battery.*

Another example can be found in the study towards realizing a museum devoted to the city of Milan (its history, culture, characteristics, development). Some of the proposed titles were:

- Milan in museum
- Milanmuseum
- Milan: Museum of the city
- Museum of the big Milan.

And some of the proposed subtitles were:

- The memory for the future
- From the past to the future
- The past for the future
- The design and memory
- From history to life
- From memory to development.

The title of an exhibition about the development of technical drawings of machines, held in Milan (1988), with the cooperation of one of the authors, was as follows (Pavan and Sabbatini 1987):

Drawings of machines (title)
Evolution of language in the evolution of technical constructions (subtitle).

The title and subtitle of another exhibition, this one devoted to the evolution of the costume jewelry industry in Casalmaggiore, a little city in Lombardy, was:

The factory of fake gold (title)
Old machines and the manufacturing of costume jewelry in Casalmaggiore (subtitle).

2.6.2 The Initial Behavior

The initial behavior of visitors to a scientific-technical museum can, in general, be expressed as follows.

- Knowledge of general culture, often, but not necessarily, good or very good;
- Knowledge of scientific-technical culture, sometimes good, but often not;
- Interest good or very good.

2.6.3 The Terminal Behavior

It is more difficult to determine the terminal behavior of visitors, because each museum is characterized by different types of terminal behavior, related directly to the contents and arguments of the museum itself. Some examples of desired terminal behaviors are presented here.

- Exhibition for the 125th anniversary of the Politecnico di Milano (1988) (aa.vv 1988)
  - *To learn about the teachings and research conducted at the Politecnico di Milano and their relative relations to the historical, industrial and environmental events from 1914 to 1963*
- Section of the Museo Nazionale della Scienza e della Tecnica (now Tecnologia) in Milan (National Science and Technology Museum) devoted to the textile industry
  - *To obtain knowledge about the most important natural and artificial textile fibers, about their specific methods of realization and about the manufacturing of the fibers for the purpose of creating fabrics and cloths*
- Museum of Didactic Activity (was proposed in 2002 and never realized)
  - *To understand the historical perspective in regards to issues within the realm of educational instruction and of the relative chronological evolution of education, for the purpose of informing the future of teaching through lessons from the past*

- Agricultural Museum
  - *To gain knowledge of the problems of agriculture, defined as land cultivation and the keeping of livestock* (Mariani et al. 2014)
- Exhibition for the 150° anniversary of the founding of the Politecnico di Milano (aa.vv 2013)
  - *To get an idea of the inventions and innovations introduced by graduates and professors of the Politecnico over the 150 years of its history*
- Exhibition about the sinking of the Titanic
  - *To learn about the events related to the sinking, including recreating the atmosphere of the trip and the dramatic moments of the disaster.*

2.6.4 The Concepts

As a first example, let us consider the exhibition for the 125th anniversary of the founding of the Politecnico di Milano (aa.vv 1988). The approach to the concept's analysis could be as follows:

(a) Institutional aspects

(a.1) engineering students
(a.2) architecture students
(a.3) foreign students
(a.4) female students
(a.5) …

(b) Teaching and research

(b.1) general situation of engineering courses
(b.2) general situation of architecture courses
(b.3) didactics in engineering: drawing
(b.4) didactics in architecture: architectural composition
(b.5) aeronautical research
(b.6) chemical research
(b.7) structural research
(b.8) …

(c) The Politecnico and social aspects

(c.1) protest again Max Abraham
(c.2) activity of certain departments in relation to war requirements
(c.3) …

(d) The role of the Politecnico in the development process

- (d.1) National Science and Technology Museum
- (d.2) Italcementi
- (d.3) Torno
- (d.4) Alfa Romeo
- (d.5) …

Another example is related to the case of the Museum of Didactic Activity. Analysis of the concepts can be detailed as follows:

(a) General principles
(b) Didactics in chronological order

- (b.1) didactic methods and means of the "first generation"
- (b.2) didactic methods and means of the "second generation"
- (b.3) didactic methods and means of the "third generation"
- (b.4) …

(c) Didactics in systematic order

- (c.1) in relation to the argument
  - (c.1.1) didactics in the literary field
  - (c.1.2) didactics in the socio-economical field
  - (c.1.3) …
- (c.2) in relation to the users
  - (c.2.1) didactics for children
  - (c.2.2) didactics for adults
  - (c.2.3) …

A further example is constituted by the agricultural museum. The conceptual analysis can be represented as follows:

- land cultivation
- land preparation
- plowing
- harrowing
- operations
- instruments
- sowing
- successive activities
- plant care
- harvesting
- utilization of land products
- livestock

- supply
- care
- obtaining products
- utilization of products.

For the exhibition related to the 150th anniversary of the founding of the Politecnico di Milano, the conceptual analysis was developed as follows. The general concepts are certain important alumni and teachers from the Politecnico, and the specific concepts are those innovations associated with those individuals and conserved at the Museo Nazionale della Scienza e della Tecnologia (National Museum of Science and Technology) (the exhibition was the result of a collaboration between the Politecnico and the above-mentioned Museum, which was itself celebrating the Sixtieth anniversary of its founding that same year). Seven "*streets*" were identified within the Museum, devoted to important *graduates* and/or teachers of the Politecnico and inhabited by the machines, devices, and instruments conserved within the Museum that were realized as a result of the efforts of the above-mentioned students/teachers.

The concepts are the names of these excellent students/teachers. In particular,

- Giuseppe Colombo
- Enrico Forlanini
- Giulio Natta
- Giovanni Battista Pirelli
- Giovanni V. Schiaparelli
- Guido Ucelli
- Francesco Vecchiacchi.

The conceptual analysis continues with the most significant innovations of the above-mentioned persons.

As an example:
Giovanni Battista Pirelli (founder of the well-known rubber company) represents the general concept, with the specific concepts being represented by three objects related to Pirelli and conserved in the Museum:

- submarine cables for telegraph use
- gum mixer
- Pirelli Tyre Type N.

Now let us look at the exhibition about the sinking of the Titanic. The above-mentioned terminal behavior can be analyzed so as to highlight the following specific concepts:

(a) Construction
(b) Technical data
(c) Technicians involved
(d) Launching

(e) Maiden voyage

(e.1) life on board
(e.2) travelers
(e.3) collision with the iceberg
(e.4) rescue operations.

2.6.5 *Museal Units*

The variety of museal units can be significant in relation to the wide possibility of integration of the above-mentioned subsections of units.

In general, in the case of scientific-technical museums, the unit can be laid out as indicated in Table 2.3, by using a synthesis of the parts presented in Sect. 2.4.

Some schematic examples are now presented.

Figure 2.6 presents the simplest of museal units: it is constituted only by a placard, with integration of written text and pictures, in which the text represents the RUL and the picture represents the critical/practical component (EG and EG′).

Figure 2.7 presents two examples in which the placard represents the RUL and the EG and EG′ are represented, respectively, by a real object on the left (of course, with comments) and by a real object accompanied by a fixed image on the right.

In Fig. 2.8, a variation is presented: the real object is also presented as a section, for the purpose of showing the internal details.

In the unit presented in Fig. 2.9, the real object is also presented through a video related to the workings or utilization of the object.

An example, related to harrowing instruments, is presented in Table 2.4.

Table 2.3 Examples of the components of terminal behavior for scientific museums

Components	Examples for scientific-technical museums
Informative	Information, generally on a placard or display, similar to the page of a book, but with great synthesis and motivational tricks
Critical	Connection of the exhibited objects to other objects and to the socio-economic environment Work environment with the aim of describing the related human and psychological situations (with particular emphasis on the emotional aspects), e.g., with light or sound effects
Practical	Overview of the exhibited objects with the aim of explaining their construction and use. Work environment and synthesis of manufacturing, with the aim of presenting a survey of technical problems "Active museum," with the possibility for the visitor of real or virtual interaction with the exhibited objects Possibility for the visitor to conduct experiments with the objects in the museum (e.g., send a message with the telegraph, listen to a phone in an old telephone box)

Fig. 2.6 RUL and (EG + EG′) integrated in a poster

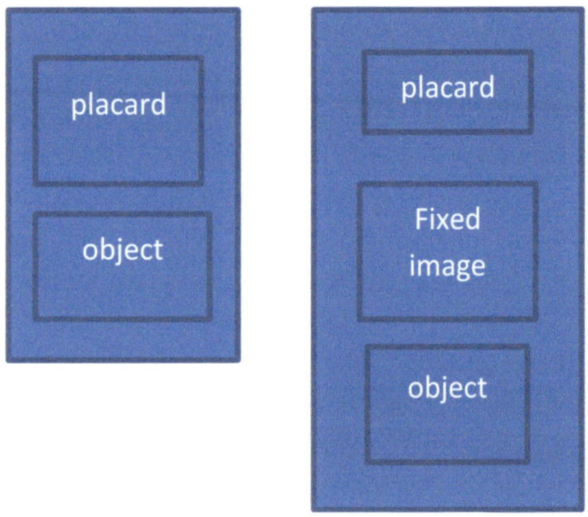

Fig. 2.7 Examples in which the placard represents the RUL and the EG and EG′ are represented by a real object on the left (of course, with comments) and by a real object accompanied by a fixed image on the right

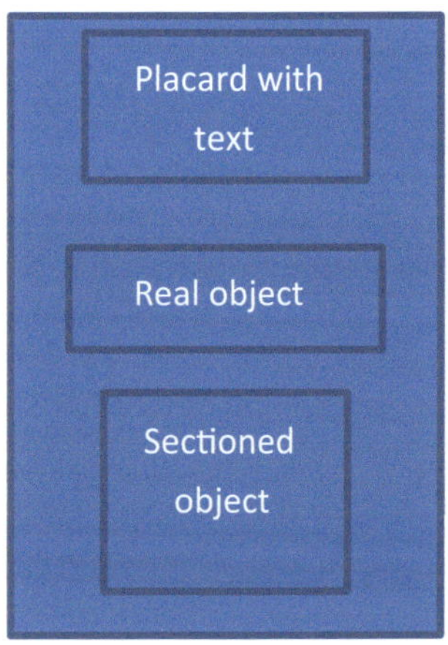

Fig. 2.8 The real object is also presented as a section, to show the internal details

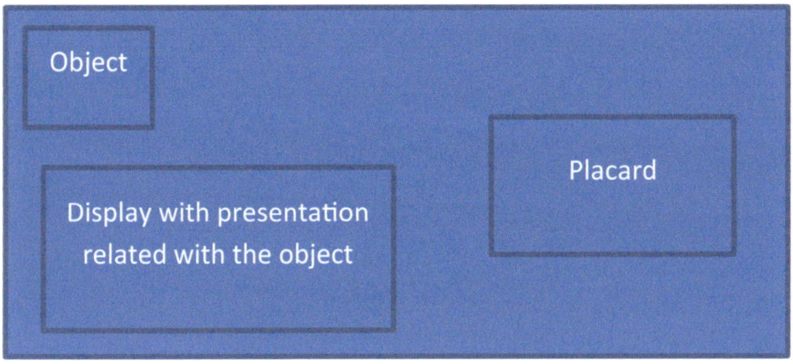

Fig. 2.9 The real object is also presented through a video related to the workings or utilization of the object

Table 2.4 Components of the terminal behavior and relative contents

Components	Contents
Informative	General description of harrowing throughout the ages
Critical	Environmental factors of harrowing and examples of the technique's utilization
Practical	Constructive solutions regarding harrowing, presented through real machines, pictures, drawings, and models Video showing examples of harrowing Living museum, to demonstrate how the harrow functions Simulation of a harrowing operation by the visitor

Figure 2.10 is related to a museal unit of "The History of Drawing" in the Agricultural Historic Museum of S. Angelo Lodigiano, near Milan.

The unit (devoted to agricultural and technical drawing in the Nineteenth Century) is constituted by a placard with text and pictures (examples of technical drawing related to agriculture), and objects (technical instruments of the Nineteenth Century).

Figures 2.11, 2.12, 2.13 and 2.14 depict placards from the exhibition for the 125th anniversary of the Politecnico di Milano (1988).

Figure 2.15 is related to a general museum unit in the technical field of machines.

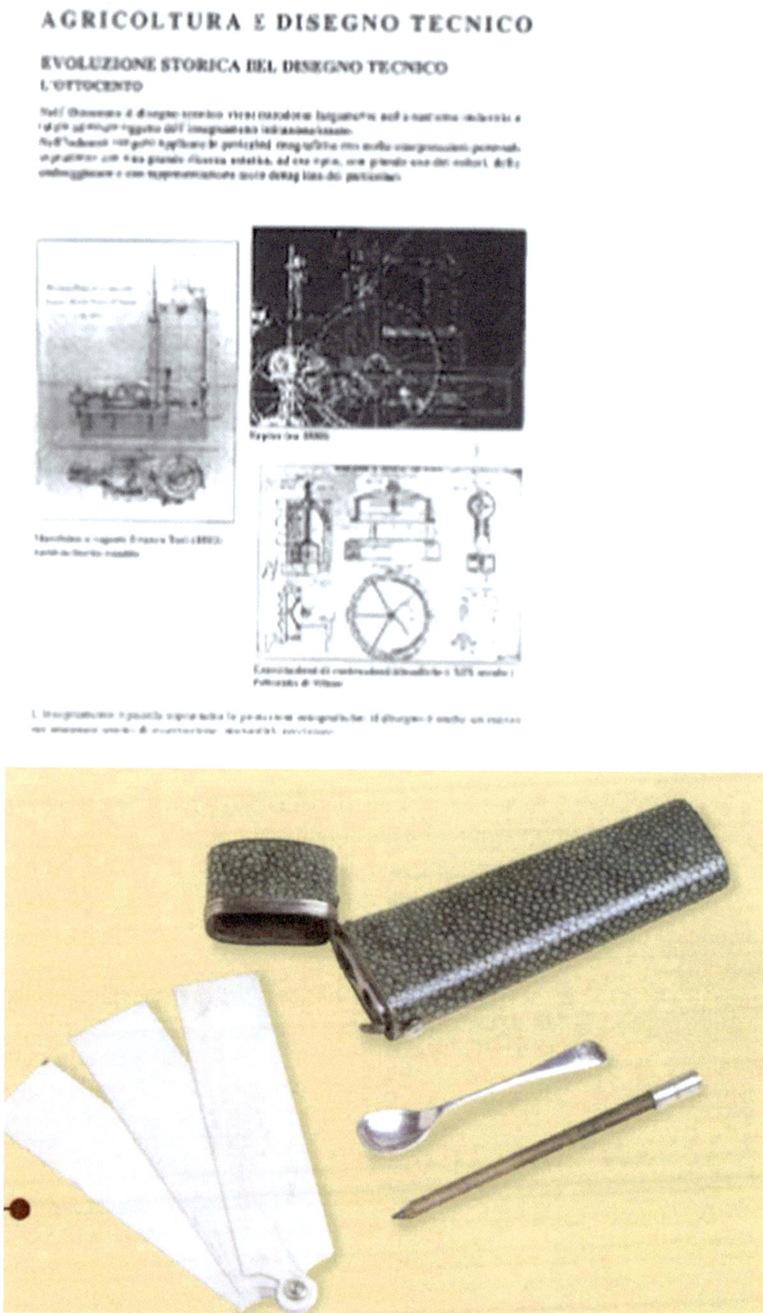

Fig. 2.10 Museal unit of "The History of Drawing" in the Lombard Museum of the History of Agriculture of S. Angelo Lodigiano, near Milan. The unit is constituted by a placard and an object

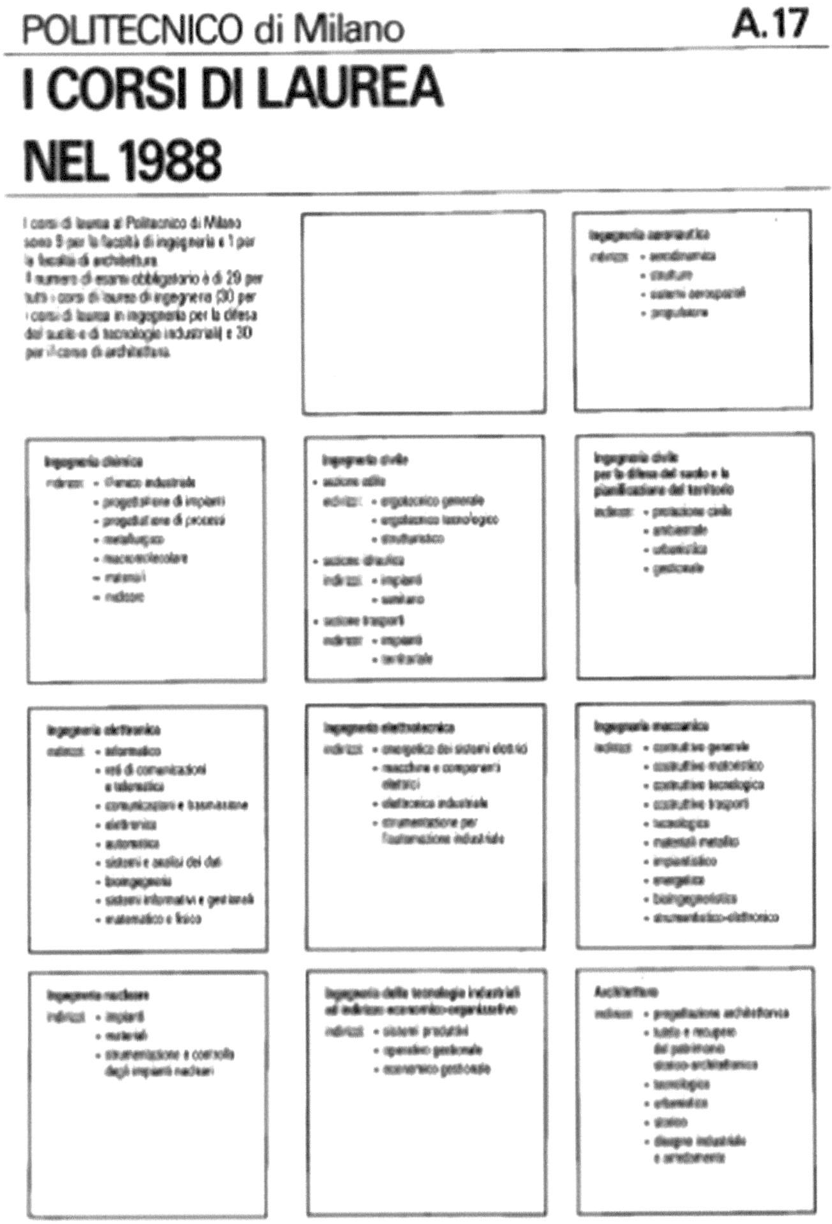

Fig. 2.11 Placard devoted to the engineering courses, constituted as a composition of tables (exhibition for the 125th anniversary of the Politecnico di Milano (1988)). The placard is constituted by an integration of text and frames

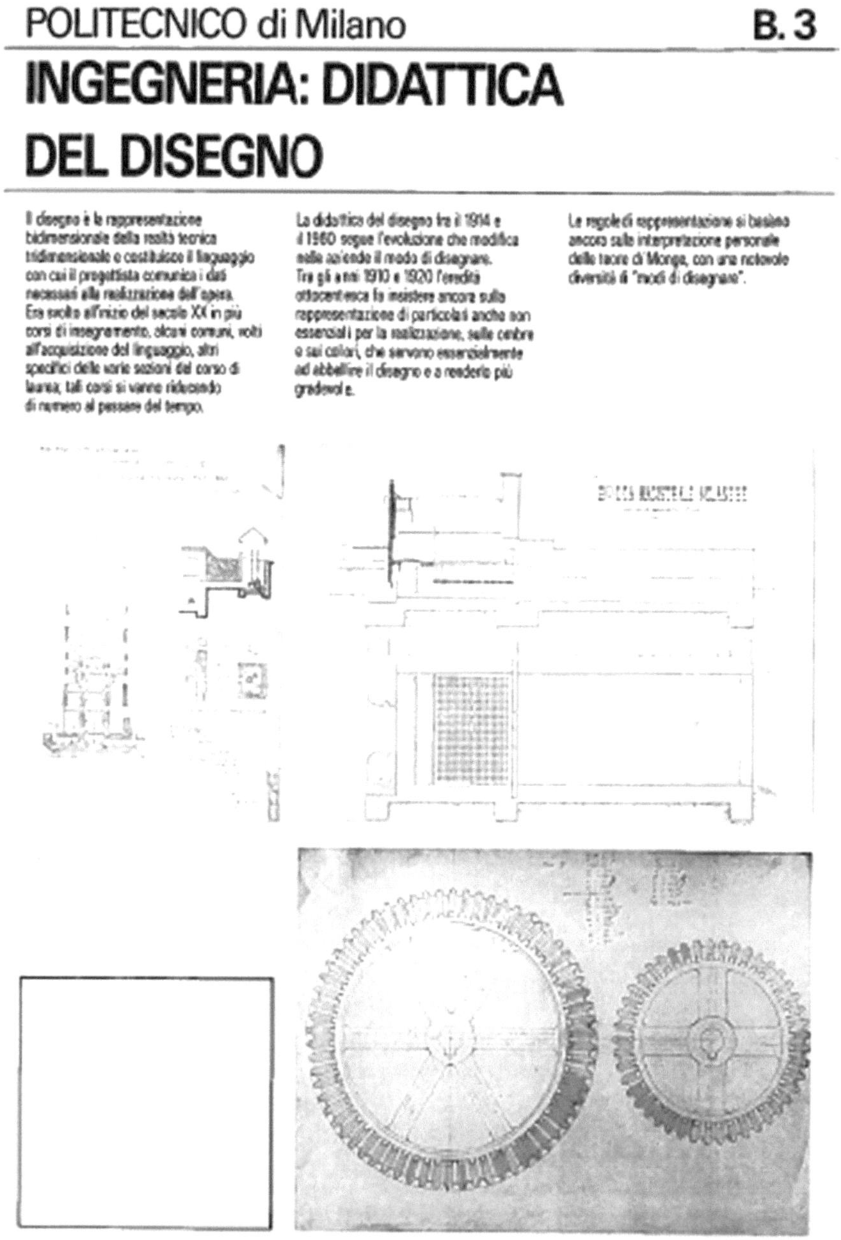

Fig. 2.12 Placard devoted to the teaching of drawing (exhibition for the 125th anniversary of the Politecnico di Milano (1988)). The placard is constituted by an integration of text and pictures, representing the exercises of "old" students

POLITECNICO di Milano B. 25

RICERCA MATEMATICA

L'Istituto è stato fondato nel 1923 da Umberto Cisotti. Cattedre (nel periodo 1923-1963): Meccanica razionale (Umberto Cisotti, fino al 1946; Bruno Finzi, dal 1948; Emilio Clauser, dal 1961); Analisi matematica (Luigi Amerio, dal 1949; Marco Cugiani, dal 1961); Geometria (Giuseppina Masotti Biggiogero, dal 1960; Ermanno Marchionna, dal 1961). Molto importante è anche il contributo di numerosi ricercatori (molti pervenuti alla cattedra in epoca successiva) e l'apporto, determinante per le ricerche in Geometria, dato da Oscar Chisini (dell'Università di Milano), il quale presso il Politecnico tiene l'incarico di Geometria dal 1928 al 1957. Oltre che per l'insegnamento di Chisini, esistono tra gli Istituti matematici dei due Atenei, notevoli connessioni, con criterio interdisciplinare evidenziato dal comune Seminario Matematico e Fisico, nonché dai corsi di incarico, tenuti per molti anni, all'Università.

Argomenti essenziali di ricerca

1. **Analisi matematica** - Analisi esistenziale e qualitativa di equazioni differenziali (legate a problemi di elettrotecnica, dinamica del punto, diffusione, propagazione delle onde, fluidodinamica) in campo lineare e non lineare: estensione ai sistemi non autonomi della nozione di punto singolare, soluzioni periodiche o quasi periodiche, riduzione a equazioni integrali,... Teorie delle funzioni quasi periodiche astratte. Analisi numerica: metodi deterministici e probabilistici,..., in relazione anche all'istituzione del Centro di calcoli numerici.
2. **Geometria** - Geometria algebrica delle linee e delle superficie: integrali abeliani dal punto di vista geometrico; singolarità, fasci e trecce che le caratterizzano; curve di Klein-Lie, teorema di Riemann-Roch, varietà di intersezione completa e varietà di diramazione... Geometria differenziale e integrale: relazioni integrali relative agli ovali e agli ovaloidi,...
3. **Meccanica razionale e Fisica-matematica** (in ambito classico e relativistico) - idrodinamica e aerodinamica: moti laminari, problema della scia, moto di singolarità idrodinamiche; fluidodinamica aleatoria,... Questioni varie di elasticità e plasti... Teoria della relatività: equazioni einsteiniane; relatività dei fenomeni irraggiamento atomico, teorie elettromagnetiche e unitarie, dina... relativistica; moto delle singolarità campi gravitazionali einsteiniani,...
4. **Storia della matematica** (ad o... essenzialmente di Arnaldo Masotti... 1934 professore di Meccanica raz... nella Facoltà di Architettura) - Inda... storico-critiche sui matematici mi... (Bonaventura Cavalieri, Maria Gaet... Agnesi, Paolo Frisi, Gabrio Piola), ... su figure cospicue in Storia della scienza: Luca Pacioli, Nicolò Coper... Nicolò Tartaglia.
5. **Seminario Matematico e Fisico ... Milano** - Istituito nel 1927 da Gian Antonio Maggi (dell'Università) e d... Umberto Cisotti, con il patrocinio e contributi congiunti dell'Università ... Politecnico. Promuove e diffonde g... studi, soprattutto mediante confere... Queste sono raccolte nei "Rendicon... Seminario Matematico e Fisico di M... che vengono scambiati con similari pubblicazioni di Università, Accade... ecc. Il Seminario costituisce così u... propria ragguardevole biblioteca ch... sede presso l'Istituto matematico d... Politecnico.

Fig. 2.13 Placard devoted to research in the mathematical field (exhibition for the 125th anniversary of the Politecnico di Milano (1988)). The RUL is very wide and the EG is constituted by two figures, representing two significant moments from the life of mathematics professors at the Politecnico

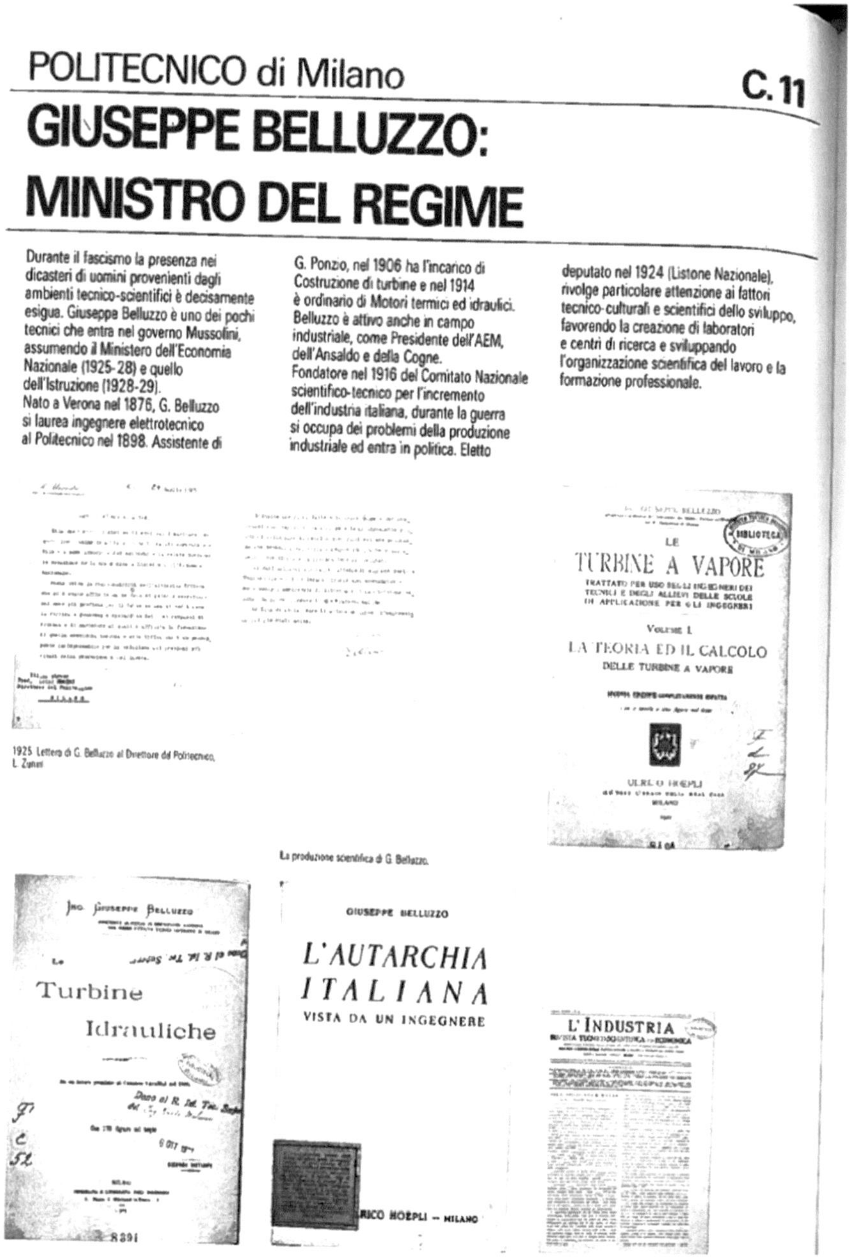

Fig. 2.14 Placard devoted to a teacher and politician (exhibition for the 125th anniversary of the Politecnico di Milano (1988)). The EG is constituted by books written by the subject

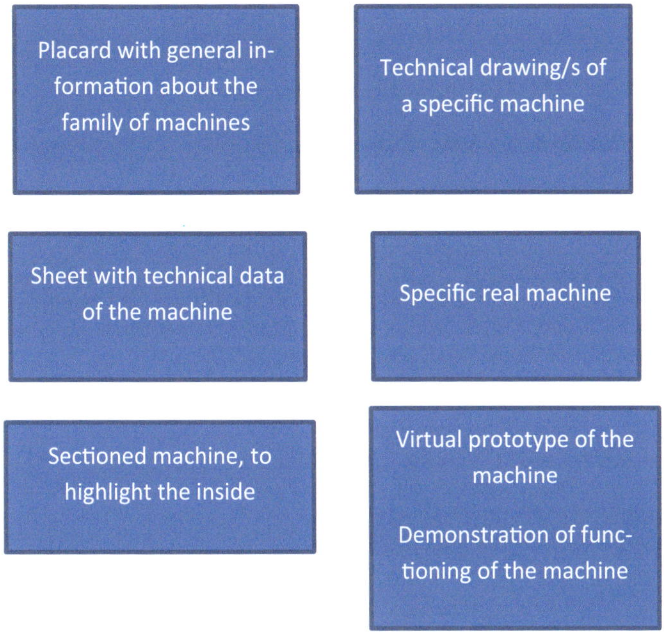

Fig. 2.15 General unit for a museum of machines

2.6.6 Museum Paths

Table 2.5 is related to the respective linear and deeper paths of a section of a museum devoted to the textile machine.

A clear delineation of the respective paths is very important, so as to prevent the visitor from suffering disorientation. In relation to this requirement, graphical indications that, in relation to the textual information, are more synthetic and independent of the native language are essential (Da Milano and Sciacchitano 2015).

Table 2.5 Example of contents of the linear and deeper paths of a section of a museum devoted to the textile machine

Components of terminal behavior	Linear path	Deeper path
Informative	Description of the obtaining of the wire	More in-depth information about the obtaining of the wire
Critical	Machines used to obtain the wire	Socio-economic environment surrounding the obtainment of the wire
Practical	Sectioned machines used to obtain the wire	Possibility of interacting with real or virtual machines used to obtain the wire

References

aa.vv (1988) Il 125° del Politecnico di Milano Cittastudi Milano

aa.vv (1999) La pila di Volta. Una scintilla lunga due secoli Mostra di cimeli dell'Istituto Lombardo Accademia di Scienze e Lettere per il bicentenario dell'invenzione Catalog of the exhibition Skira

aa.vv (2013) Made in Polimi 1863–2013 Electa

Biggioggero GF, Rovida E (1977) Problems in the teaching of mechanics. Eur J Eng Educ 2 (1):277–290

Biggioggero GF, Calabrò S, Gorla C, Menzio G, Rovida E (2003) The virtual museum of machine elements as mean for the teaching in the drawing field. In: XIIIADM—XV INGEGRAF international conference "tools and methods evolution in engineering design" Cassino, 3 June 2003, Napoli, 4 and 6 June, 2003, Salerno, 5 June 2003

Da Milano C, Sciacchitano E (2015) Linee guida per la comunicazione nei musei: segnaletica interna, didascalie, pannelli Quaderni della valorizzazione MiBACT. Ministero dei beni e delle attività culturali e del turismo, Roma

Mariani L, Forni G, Rovida E, Failla O (2014) Guida alla visita ragionata Museo Lombardo di Storia dell'Agricoltura (Lombard Museum of the History of Agriculture). Available on https://sites.google.com/site/storiagricoltura/museo. Last visit 12 Mar 2017

Pavan L, Sabbatini A (1987) Disegni di Macchine. Evoluzione di un linguaggio nello sviluppo della tecnica Catalog of the exhibition, Grafiche Editoriali Artistiche Pordenonesi, Pordenone

Shannon CE, Weaver W (1962) The mathematical theory of communication. University of Illinois Press, Urbana, IL

Chapter 3
General Principles About Museums and Exhibitions

3.1 Introduction

In this chapter, some general considerations about museums and exhibitions are presented. First, we observe that museums and exhibitions present many reciprocal analogies. Both have, as a fundamental aim, the diffusion of the culture through exposition of collections of objects. Among their most principal differences, museums are permanent, while exhibitions are temporary. Additionally, in most if not all cases, museums are more general and exhibitions more specific. This difference often plays a role in the structure of the two institutions. The exhibition, for its specificity and impermanence, is generally characterized by a great synthesis, bringing together, for the opening period, objects generally located in different, and often distant, museums. It also not infrequently allows people to see materials from the museums' vaults that are normally difficult (or even impossible) to observe, while its transient nature can cast the exhibited objects in a more dynamic light.

It is necessary to observe that, as highlighted in the previous chapters, the emotional aspect of the presented objects is very important.

Their respective expositive languages, therefore, should be different: more considered in the case of museums, more dynamic in the case of exhibitions.

At any rate, most of the following considerations can be applied to both institutions.

3.2 Objectives

The general objective of a museum (in general, and particularly for a scientific-technical museum) can be defined as the creation and diffusion of culture through conservation and exposition. The museum, therefore, is not only a site where certain objects are exhibited, but must provide an educational message, by

A. Rovetta and E. Rovida, *Scientific Knowledge Communication in Museums*, Research for Development, https://doi.org/10.1007/978-3-319-68330-0_3

arousing interest in science and technology, through curiosity and wonder above all. In addition, it bears repeating that the exhibited objects should be presented in such a way that highlights their emotional aspects. The museum, therefore, not only collects and presents objects, but also, above all, the scientific-technical knowledge that produced them.

The museum also serves the very important function of orienting young people: the possibility of observing and studying a work environment and production processes from a synthetic viewpoint provides a valuable opportunity to gain cultural and professional experience.

3.3 Functions

Starting from the above-mentioned objectives, and through consideration of the definition of the museum proposed by ICOM (Sect. 1.3), the following functions of a museum, particularly of the scientific-technical type, can be recognized:

- Collection of objects constituting a scientific-technical heritage;
- Conservation of the above-mentioned objects, i.e., restoration of such objects, with the aim of placing them within conditions to be observed and studied, using all necessary expedients so that the objects themselves remain preserved for future generations;
- Exposition, i.e., presenting the objects in locations and conditions suitable for observation and study; in the exposition, the rational aspects must be taken into account, but so must the emotional;
- Research, i.e., study of the exhibited objects (as well as those in storage), with the aim of obtaining original knowledge about different aspects of the objects themselves;
- Didactics, i.e., transmission of knowledge that begins with the exhibited objects. From this point of view, it is important to distinguish the three fundamental aspects of knowledge, i.e., the informative, the critical and the practical.

3.4 General Configuration

The scientific-technical museum may present a great variety of types and configurations. But among such variety, some general considerations should be made, in relation to the following general parts of a museum.

3.4.1 Location

The individuation and choice of the location is a very important phase. In some cases, the choice is mandated by certain pre-existing considerations (e.g., historical exactitude): in such a case, there is no choice problem. When it is, however, possible to choose, the following points of view can be of interest.

- Presence of constraint: for example, the adjacency of airports or roads with heavy traffic can be negative because of the noise and vibrations, but positive because they can facilitate the connection with the museum; the adjacency of industries, however, is always negative;
- Suitable buildings: for example, an old decommissioned station could be a good location for a railway museum, a decommissioned hangar or aviation industry building a good location for an airplane museum. In general, the availability of an historical monumental building can be interesting, because the *container* itself can be a museum;
- Links to cultural relevance of the environment: for example, an archeological museum can suitably be located within an archeological environment;
- Links with the urban character of the environment: an interesting museum can be useful for qualification of the urban environment in which it is built, like the Museè de la Villette in Paris;
- Dimensions of the building, which must be in strict relation with the dimensions of the collections to be exhibited;
- Immediacy of availability, linked to the necessity to restructure the location for practical purposes;
- Ease of connection, e.g., with the transportation that will be used to reach the museum, or with the locations of other related museums;
- Clear indications as to how to reach the museums (itineraries, transportation means, and so on).

Some interesting examples of museums with particularly apt locations can be read about in Muehlenthaler (1979), Allégret (1992) and Hoffman (2016).

3.4.2 Introduction

The introduction can be defined as the part of the museum (a section, a single room, or, even more simply, a portion of a room or just a placard) where all of the necessary (or at least useful) information is presented to the visitor, with the aim of optimizing the visit. It is very important that the visitor be put in the center of the visit. At the entrance to the museum, an object with general interest or a general placard can be presented. The object would, of course, be something significant to the collection of the particular museum: an airplane, a railway vehicle, a car, an agricultural machine, etc., accompanied by explanatory placards. The aim of these objects, fundamentally, is to stimulate the emotion and interest of the visitors. In regard to placards, examples could be a diagram showing the fundamental steps of the historical evolution of a machine (for a scientific-technical museum) or a geographical world map with indications of the origin of certain significant edible plants (for an agricultural museum).

In principle, the introduction can be constituted of the following parts:

- **Characteristic object** related to the collections;
- **Objectives and mission** of the museum, i.e., why the presented objects are important, the upgrade in knowledge that the visitor can expect to reach through visiting the museum (defined in Chap. 2 as the "terminal behavior");
- **"Index,"** i.e., a list of the sections of the museum, perhaps presented through a characteristic object, representative of the section;
- **Criteria of the visit**, i.e., suggestion for the visitor of an optimal sequence in which to view the different sections of the museum;
- **Timetable of the events** organized by the museum (e.g., courses, conferences, laboratories, specific exhibitions);
- **History of the museum:** many museums were founded many years ago (in some cases, centuries). In these cases, a placard with some fundamental information about the history of the museum is interesting. Linked to this, some books related to the museum or to the exhibited materials could be presented;
- **Information about the building where the museum is located:** many museums are located in beautiful historical buildings. Examples include the Museo Nazionale della Scienza e della Tecnologia (National Museum of Science and Technology) "Leonardo da Vinci" in Milan, which is located in an Olivetan monastery of the Sixteenth Century, and the Museo di Storia dell'Agricoltura Lombarda (Museum of the History of Lombard Agriculture) in S. Angelo Lodigiano, near Milan, which is located in a Visconti castle;
- **Synthesis (e.g., through placards) of the exhibition** and events organized by and in the Museum;
- **Survey about museums** devoted to an analog argument that could be interesting for some visitors; this survey can be presented through placards, and also, of course, through video;
- **Video, in many languages, with explanations** by the curator of the museum;
- **Video or placard with photos and biographical information** about the founder of the museum;
- **Tactile floor plan** for blind or visually-impaired visitors.

Much of the above-mentioned information can be presented on a placard with big dimensions, disposed horizontally on the floor and, obviously, realized in materials that pose no danger of damage. The information should also, of course, be presented on the museum's website.

3.4.3 Exposition

The exposition can be defined as the fundamental part of the museum, where the material is presented for observation and study by the visitors. The collections are sets of objects that have been formed over time through different means (purchases, donations, bequests), and are among the main elements that characterize the museum.

The exposition means, in general, conservation of the materials (i.e., maintaining the materials in a good condition of efficiency, necessary for correct observation) and teaching (i.e., utilization of the materials to transmit a message).

The exposition requires an accurate phase of design and realization of the museum, starting from the consideration that the museum is a form of technical communication (see Chap. 2). The result of the design of the technical communication corresponding to the museum can be defined as an integration of objects, explanatory texts and related documents, placards, displays that may include deeper levels of information, spoken commentary, and musical background, as will be explained in the subsequent chapters.

During the design phase of a museum, certain considerations can be made, such as the following:

- Disciplines to be represented in the museum;
- Collections to be considered, such as real objects, real or virtual models, images, etc.;
- Chronological limits of the exposition;
- Geographical limits of the exposition;
- Possible limits linked, e.g., to specific persons, places, or events;
- Definition of the qualitative standard of the collections, such as, e.g., national importance or aesthetic level;
- Aims of the exposition, e.g., representative of ideas, concepts, schemas, associations linked to specific situations, places or persons.

After the design of the museum, the mounting is the translation of the design into the product, i.e., in the physical museum. In this phase, as an example, the following steps can be recognized:

- planimetric development (museum spaces and paths);
- vertical development (form, dimensions, material, colors);
- lighting;
- protection of the exhibited materials (windows, barriers);
- written and graphical communications (text, captions, placards, indications);
- virtual communications.

Some authors distinguish:

- study of the spaces (plants, vertical development, massing);
- study of the speech (exposed materials, text, placards, dioramas).

Table 3.1 Example of a simple questionnaire for compiling the opinions of visitors

Name		
Email (if answer is required)		
Comments/observations about the museum		
Best aspects		
Worst aspects		
Is an answer required?	Yes	No

Of course, during the design phase of the museum, it should be useful to consider that the exhibited materials (or the most significant among them) must be available on the museum's website.

3.4.4 *Conclusion*

It could be interesting to have, at the end of the visit, a section or a room featuring the following items:

- A short summary related to the sections of the museum, with a reminder of the most significant piece (2D or 3D model) in each section;
- A digital catalog of the museum, with the possibility of repeating the visit in virtual form and seeing the most interesting pieces again;
- A map with indications of other nearby museums;
- Questionnaires, with the aim of compiling visitor opinions (an example is shown in Table 3.1);
- A bookshop;
- A library with reading stations;
- Location with 3D printing machines and the possibility of realizing a 3D model of exhibited objects.

Information about the design and realization of museums can be found in Arestizàbal and Piva (1991) and Dal Co and Muirhead (1997).

3.5 Examples

3.5.1 *Exhibition of the History of Drawing of Machines (Pavan and Sabbatini 1987)*

The ADM (Associazione Nazionale Disegno di Macchine) (Italian Association of Drawing and Design of Machines), during their Conference in 1986 (in Udine) and, successively, in 1987–88 (in Milan), organized two exhibitions with the aim of explaining and presenting the history and evolution of machine drawings.

The steps of the process were as follows:

- Constitution of the scientific committee. The scientific committee was made up of teachers of Machine Design and Machine Drawing from the Politecnico di Milano, the Politecnico di Torino and the Universities of Padova, Udine and Trieste
- Individuation of the materials to be exposed. The materials to be exhibited were individuated according to the knowledge and through the direct contact of the members of the scientific committee. Contacts and research were coordinated according to the Institutions as delineated in Table 3.2.
- Selection of the material to be exhibited. After individuation of the materials, a selection was made in regard to the objectives of the exhibition. With reference to Chap. 2, such objectives, defined as the "terminal behavior," are expressed as "present to persons, with good general culture and without very deep technical culture, the historical evolution of the representation of machines and mechanisms." It is easy to recognize that, in the above-mentioned definition of the objectives, the prerequisites, expressed in Chap. 2 as the "initial behavior," are also implicit. Starting from this definition, the selection of the materials to be exhibited was realized, by choice of the most important pieces, with reference to the representativeness of the pieces themselves, taking into account the desired terminal behavior. The selected materials belonged to the following categories
- Technical drawings of machines, parts of machines, and industrial plants
- Some real mechanical elements, such as physical examples related to the exhibited drawings.

Table 3.2 Example of the Institutions interested in the realization of the exhibition relative to the history of machine drawings

Categories of Institution	Institutions
Museums	Museo Nazionale della Scienza e della Tecnica (now "Tecnologia") (National Science and Technology Museum) "Leonardo da Vinci" in Milan
Libraries	Central Library of the Politecnico di Milano Istituto Lombardo Accademia di Scienze e Lettere (Lombard Institute Academy of Science and Literature) in Milan
Industries	Alfa Romeo Franco Tosi ABB (Asea Brown Boveri) Riva Calzoni
Associations	Alumni Association of the Politecnico di Milano UNI (Ente Nazionale Italiano di Unificazione) (Italian Standardization Institute)
Private collections	Collection of drawing instruments of Edoardo Rovida

- Books, related to the teaching and/or professional activity, of machine drawings
- Drawing instruments.

For each piece, the authorization of the owners was verified, and a cataloging sheet was compiled. Such a cataloging sheet should contain the following information.

General information

- Name of the exhibition
- Name of the compiler
- Number of the sheet
- Date of the survey
- Object (drawing, book, mechanical element, and so on).

Formal aspect

- Author/authors
- Dimensions
- Material
- Execution
- Title/name
- State of conservation
- Formal interest.

Aspects related to the represented object

- Description
- Function
- Technical peculiarities.

Notes

- Individuation of the location
- The fundamental requirements of the location are as follows.
  - availability for the required period
  - spaces congruent with the exhibition's requirements
  - exhibition facilities, such as showcases, supports for placards
  - custodial staff
  - fire prevention systems
  - ease of loading/unloading.

Based on the above-mentioned requirements, the chosen locations were the Museo della Città (City Museum) in Udine (for the 1986 edition) and the Museo Nazionale della Scienza e della Tecnica (now Tecnologia) (National Science and Technique Museum, (now Technology)) "Leonardo da Vinci" in Milan (for the 1987–88 edition).

- Choice and determination of the spaces in which the materials are to be exhibited. An accurate relief of the space is necessary, so as to optimize the

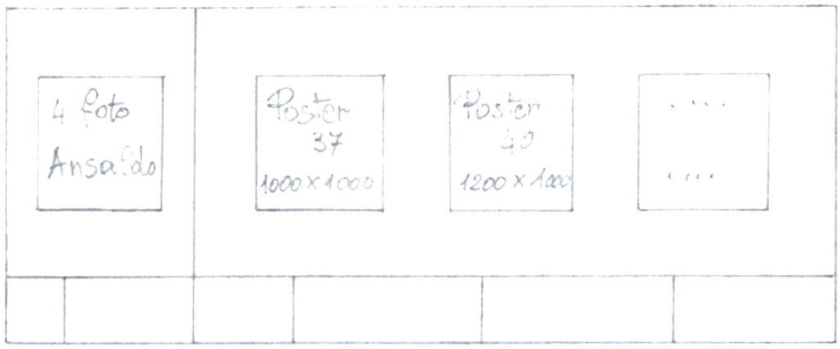

Fig. 3.1 Example of a relief of a vertical wall

Fig. 3.2 Examples of an exposed object

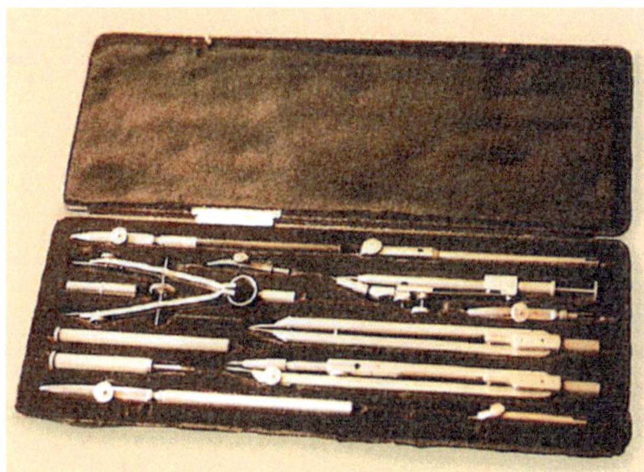

exposition of the drawings and objects. Figure 3.1 presents an example of a relief of a vertical wall.

- Mounting of the exhibition. The mounting of the exhibition was a heavy and challenging activity. In Udine, 179 pieces were exhibited, and the time required to set it all up was 150 h/man, while in Milan, the exhibited pieces numbered about 200, and the time requirement was 200 h/man.

Figure 3.2 depicts an exhibited object.

3.5.2 Museum of Machine Design (De Alberti and Rovida 1999)

The idea of a museum dedicated to machine design was initially conceived by the Mechanical Department, in support, not surprisingly, of the teaching of the course "Machine Design." The initial behavior of the students relative to the above-mentioned course was evaluated as being some small degree of knowledge about the drawing standards and some basic knowledge about the materials and the elements of machines.

The desired terminal behavior could be expressed as follows: to *reach some degree of knowledge about the representation of certain elements of machines, with individuation of the functional surfaces, learning about the tolerances, the utilization of materials, particularly in relation to the choice of material congruent with the form and the function of a given element of a machine, and the design expedients of elements of machines in relation to manufacturing. In addition, it was hoped that visitors would absorb some information about the historical evolution of mechanical construction and the teaching of mechanics.*

Analysis of the terminal behavior allows us to highlight the following concepts:

- Complements of representation
- Tolerances
- Utilization of materials
- Design expedients
- Examples of constructive solutions.

The museum's units were realized through machine elements, sometimes sectioned off or presented with didactic expedients, explicative text and placards. In some cases, active participation by the students was possible: for example, handling the drive of a mechanism or machine elements to get a sense of the functioning, or holding in one's hands similar machine elements constituted from different materials, so as to, for example, feel the different weights.

In addition, some documents related to the historical evolution of the teaching of machine design were exhibited.

In Fig. 3.3, some examples of didactic notes are represented. These notes, related to the Machine Design course, were written by a student (academic year 1922–1923).

In Fig. 3.4, some examples of a thickness calliper (circa the 1930s) are represented. Such instruments bear witness to the measurement operations in regard to the elements of machines.

Figure 3.5 depicts a model of a mechanism.

Figure 3.6 is related to different constructive solutions in regard to connecting roads for internal combustion engines.

Figure 3.7 depicts a sectioned valve, highlighting its internal configuration.

It is easy to recognize the validity of the above-mentioned and presented objects in the teaching of Mechanics. Of course, today, virtual models are most often utilized, but the Author's opinion is that the real models, and the possibility of operating them, can be a useful integration with the virtual reality.

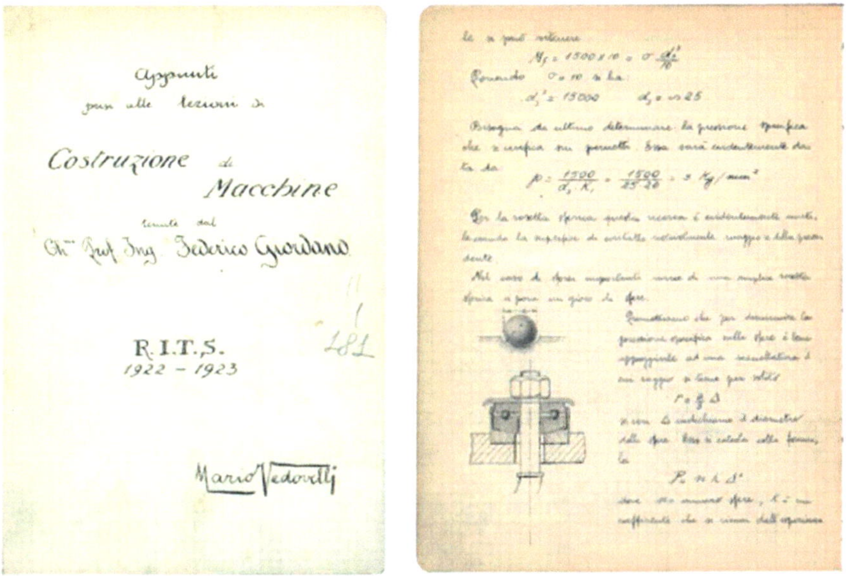

Fig. 3.3 Didactic notes related to the Machine Design course, written by a student (academic year 1922–1923)

3.5.3 Museum of Drawing (Brunetti and Rovida 2001, 2005)

The availability of a collection of historical (Eighteenth and Nineteenth Century) drawing instruments was the initial impetus for the founding of a Drawing Museum, to be realized as a specific section of the Lombard Museum of the History of Agriculture in S. Angelo Lodigiano. A drawing museum as a subsection of an agricultural museum appears to be very opportune, because there are many links between technical drawing and agriculture, e.g., drawing instruments are utilized during surveying, and all agricultural constructions, such as machines, buildings, and hydraulic plants, require technical drawings for their realization. From the above-mentioned considerations, the initial and terminal behaviors (respectively, requirements and objectives) of the communication become as indicated in Table 3.3.

The concepts determined by an analysis of the desired terminal behavior are as follows:

- General principles
  - graphical communication
  - technical drawing
  - kinds of technical drawing

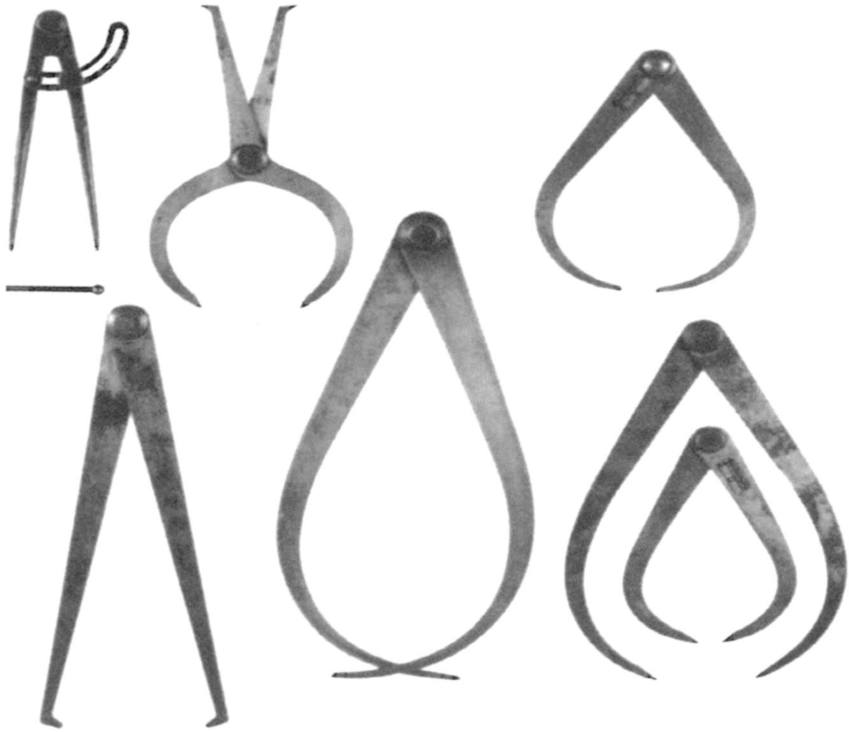

Fig. 3.4 Thickness callipers (1930s)

- Historical evolution
  - Ancient Times
  - Middle Ages
  - Renaissance
  - Eighteenth Century.

Monge and the theory of projections
Diderot e d'Alembert and the Encyclopedists

- Nineteenth Century

  initiation of industry
  initiation of drawing as an academic discipline

- Twentieth Century

standardization
development of the industry
development of the academic discipline
computer application

Fig. 3.5 Example of a mechanism

The museum's units (the parts of the museum devoted to each concept) are, in general, constituted by texts (the informative part, or rule) and pictures and objects (the critical-applicative part, or example). In the Fig. 3.8 an example by the representative placard in an integration of text (the informative part) explaining the four pictures taken from Monge's book, two applications of the projections and an example of a technical drawing of the above-mentioned picture (the applicative part). The applicative part is then completed with inclusion of a drawing instrument.

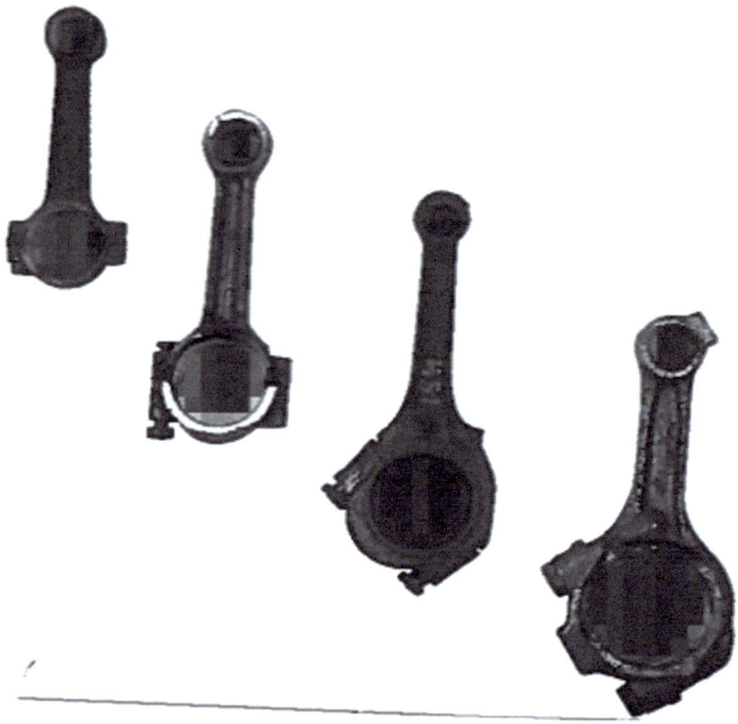

Fig. 3.6 Examples of constructive solutions of connecting rods

Fig. 3.7 Example of a sectioned valve

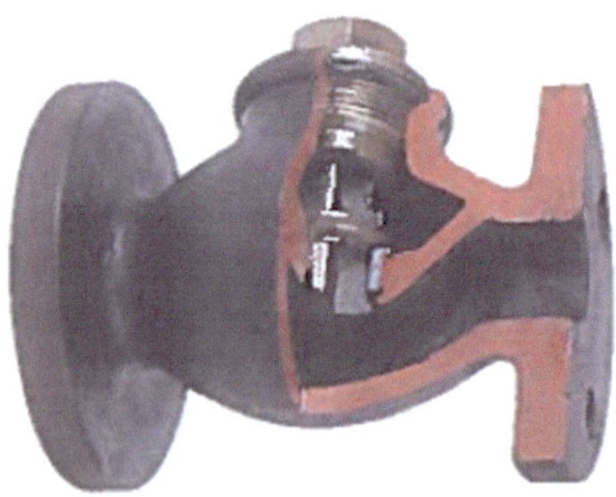

Table 3.3 Initial and terminal behaviors assumed for the realization of the museum of drawing

Step	Content
Initial behavior	Good general knowledge of culture Knowledge of technical culture not necessarily present Very good interest
Terminal behavior	General principles and historical evolution of the representation of technical products with particular reference to agriculture

3.5.4 Approach for a Wine Museum

A wine museum could be organized with reference to the following parts.

1. Introduction
   - characteristic subject of appeal (this could be a vat or an antique wine press)
   - map of the wines made in the areas around the museum
   - panel describing the museum's objectives and general information about it
   - video with general information (perhaps the Curator of the Museum talks about how the idea was born, what the key features of the museum are, what the objectives are, and so on)
   - multimedia presentation of the winery and its wines
2. Exhibition
   The exhibition would be an integration of various media, such as:
   - placard describing the history of wine
   - explanatory placards of the "working cycle" that leads from the grapes to the wine
   - explanatory placards about the use of wine (for example, types of wine and their pairings with different dishes)
   - significant objects for each phase, from the processing cycle to the use of the wine
   - archeological artifacts, tools and technical objects involved in winemaking
   - documents on the role of the wine in life and culture
   - drawings and pictures related to the wine
   - captions of each object (including name, age, brief explanation)
   - one or more explanatory videos of winemaking
   - sound commentary (such as traditional folk songs of vintners, recitations of poems with wine as their subject)
3. Conclusion
   - summary panel with the most significant information acquired by the visitor
   - bookshop that sells books and publications about wine and related food topics
   - wine shop that sells wines and related products

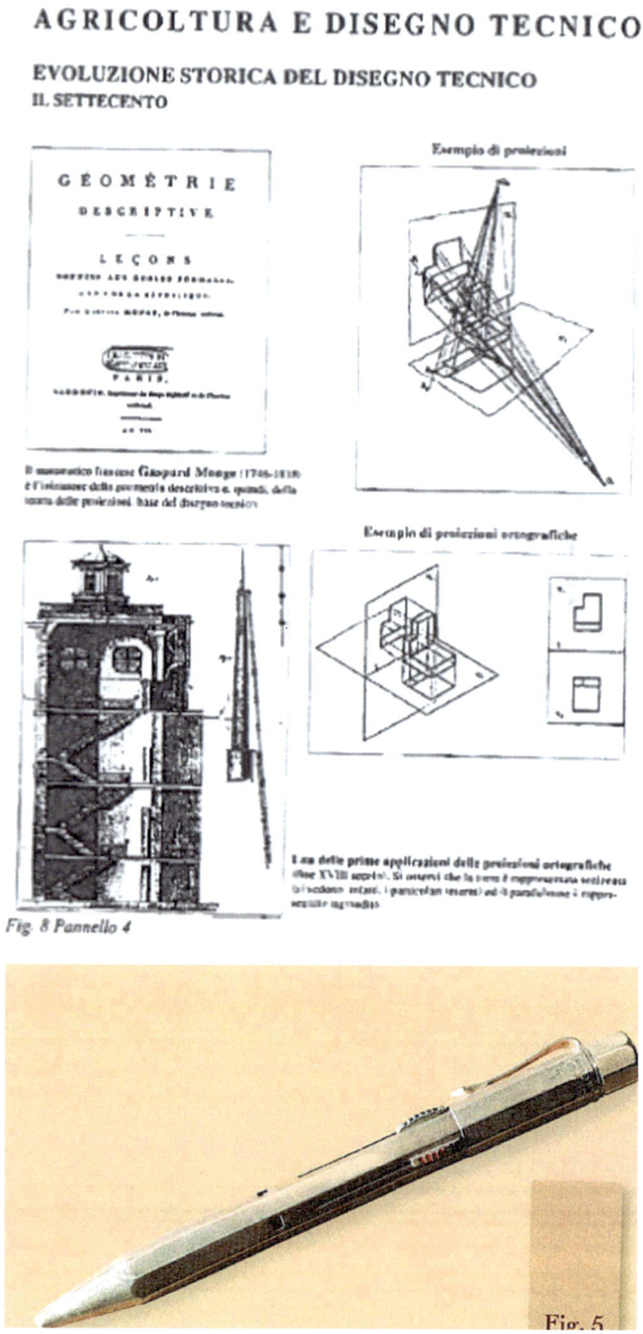

Fig. 3.8 Part of the museal unit consisting of a placard with information and examples, and a drawing instrument

- the emotional choice: wine selection based on your mood and personality, identifying combinations of the emotions of the taster and the smells, tastes and feelings aroused by the wine cellar

3.5.5 *Approach for a Museum Devoted to a Person*

One of the authors is involved in the organization of a little museum devoted to Guido da Vigevano, an engineer and physician from the Middle Ages, who was born in Vigevano, a town near Milan. Such a museum could be organized with reference to the following parts.

- Introduction
  - video describing how Guido first conceived of his studies
  - video showing reconstructions of Guido's wagon
  - panel describing the objectives of the exhibition.
- Exposition
  - Panels related to: Vigevano in the Middle Ages, the life of Guido, Guido the physician, Guido the engineer, the wagon
  - Showcase: Guido's wagon with photos reconstructing the various stages of construction
  - Video film about the wagon
  - Soundtrack of some historical film, for example, featuring the noise of a medieval battle
  - Recitation of epic historical and/or medieval poetry.
- Conclusion
  - free distribution of documentation about Guido and the city
  - sale of gadgets

References

Allégret L (1992) Musée Edition du Moniteur, Paris
Arestizàbal I, Piva A (1991) Musei in trasformazione Mazzotta Milano
Brunetti R, Rovida E (2001) Proposal of a drawing museum XII ADM (Associazione Nazionale Disegno di Macchine) (Italian Machine Drawing and Design Association) International Conference, Rimini, Italy September 5th–7th, 2001
Brunetti R, Rovida E (2005) About the institution of a drawing museum Congreso Internacional Conjunto XVII INGEGRAPH—XV ADM (Associazione Nazionale Disegno di Macchine) (Italian Machine Drawing and Design Association), Siviglia, 1st–3rd September 2005
Dal Co F, Muirhead T (1997) I musei di James Stirling, Michael Wilford and Associates Electa Milano

De Alberti L, Rovida E (1999) Historical Heritage of Mechanics Department Telesma Milano
Hoffman HW (2016) Construction and Design Manual. Museum Buildings DM Publisher
Muehlenthaler B (1979) Kleines Handbuch der Konservierungstechnik Verlag Paul Haupt Bern
Pavan L, Sabbatini A (1987) Disegni di Macchine. Evoluzione di un linguaggio nello sviluppo della tecnica Catalog of the exhibition, Grafiche Artistiche Editoriali Pordenonesi Pordenone

Chapter 4
Placards

4.1 General Principles

The placard can be defined as a means of communication constituted by a table, generally of large dimensions, with some alphanumeric and graphical information. Due to the large dimensions, the placard can be read by many people at the same time and also from a certain distance. To achieve such a result, the placard must also very perceptible, through use of a proper ratio between text and pictures: in other words, the placard should have a high percentage of graphical communication. Very often, in museums and exhibitions, placards, as a consequence of a valid aesthetic study, are quite pleasing to the eye.

By considering the theories of scientific-technical communication (as explained in Chap. 2), the placard is a communication unit (or part of a communication unit) devoted to the transmission of a concept. From this point of view, the placard can, in general, take on one of three fundamental configurations:

- Placard constituted only by a rule. In this case, the placard is constituted only by text. Such a configuration, however, is inadvisable, because to read a long piece of text, particularly one that is perhaps not very synthetic, can often be unpleasant, having a negative effect on attention and motivation. In addition, emotional involvement tends to be very low.
- Placard constituted only by an example (or part of an example). In this case, the placard is constituted only by a picture (or pictures): the rule is intended as presented in a previous placard or, if the terminal behavior of the concept related to the placard is very simple, the rule can be considered to be implied and, consequently, obvious, or, in other cases, the rule corresponds to the captions of the pictures. In many cases, the communication unit is completed by inclusion of some objects, with commentary, or by a video: in such cases, the commentary on the objects, and often a part of the spoken commentary from the video, can be regarded as a rule.

A. Rovetta and E. Rovida, *Scientific Knowledge Communication in Museums*, Research for Development, https://doi.org/10.1007/978-3-319-68330-0_4

- Placard constituted by a rule and an example. This case is, of course, a combination of previous cases. The placard is constituted by text (the informative part of the communication unit, or rule) accompanied by a picture/pictures that both explain the text and link it to a concrete situation. In many cases, the communication unit in this configuration is completed by some objects, with commentary, or by a video: in such cases, the commentary on the objects, and often a part of the spoken commentary from the video, can be regarded as an upgrade of the text written on the placard.

4.2 Design

The phase of design of a placard requires some steps.

- individuation of the concept. Due to the low density of information that characterizes the placard, it is useful for the concept being transmitted to have a low level of terminal behavior. Of course, such terminal behavior should be carefully evaluated in relation to the level of the users and to the total number of placards
- definition of the informative part (rule).The rule should be expressed through text. Such text must be very synthetic, almost lapidary, without unnecessary words that add nothing to the transmission of the message, may cause distraction among the users and essentially represent unproductive occupation of space on the placard. The placard, therefore, must be very different from a page in a book!
- definition of the applicative part (example). The example in a placard can take various forms and must be chosen very carefully, with the aim of explaining the rule effectively and without distractions and distortions for the users. It can take the form of drawings, photos, or images. In some cases, the example can be text, e.g., a document, the cover of a book, a letter. In other cases, the example can be constituted by objects, of course, only those small enough to adhere to the plane of the placard
- composition of the placard. The rule and the example/s can be combined in various ways on the surface of the placard. In this phase, a grid can be useful as a reference for the above-mentioned parts. Figure 4.1 is related to the grid utilized for the placard of the exhibition A *two-century-long spark: A commemorative exhibition of 200 years of Volta's battery*. The words in the figure translate as follows: testo (text), immagine (picture), dida (caption), una scintilla lunga due secoli (A two century-long spark).

In the grid, it is easy to recognize the distribution of available space into text (rule) and images with captions (examples). In many cases, as in the above-mentioned figure, it can be useful to place such information as the wording related to the title and the date of the exhibition at the edges of the placard. This is particularly useful in cases of reproductions of a sole placard: it is easy to recognize the exhibition to which the placard refers.

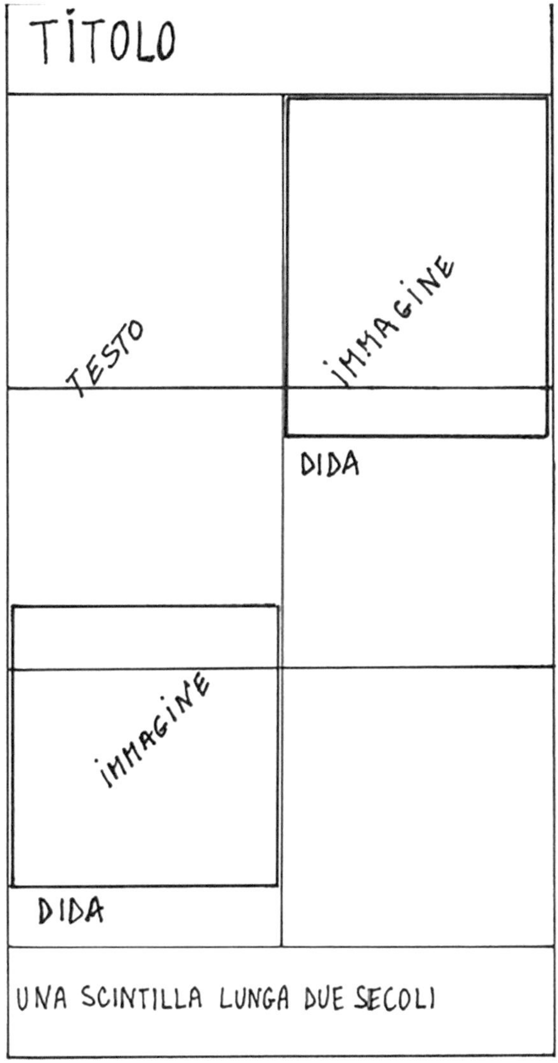

Fig. 4.1 Grid utilized for the poster of the exhibition a two century-long spark: a commemorative exhibition of 200 years of Volta's battery

Figure 4.2 is related to the exhibition for the 125th anniversary of the founding of the Politecnico di Milano. Similarly, the words translate as follows: testo (text), immagini (pictures), titolo (title).

In some cases, it can be useful to have an abstract of the placard in Braille characters.

The placard can also be utilized to present the original results of researches in scientific congresses. Many of the above-mentioned considerations continue to be valid. The configuration of the placard is not very different. Figure 4.3 is related to an example of this type.

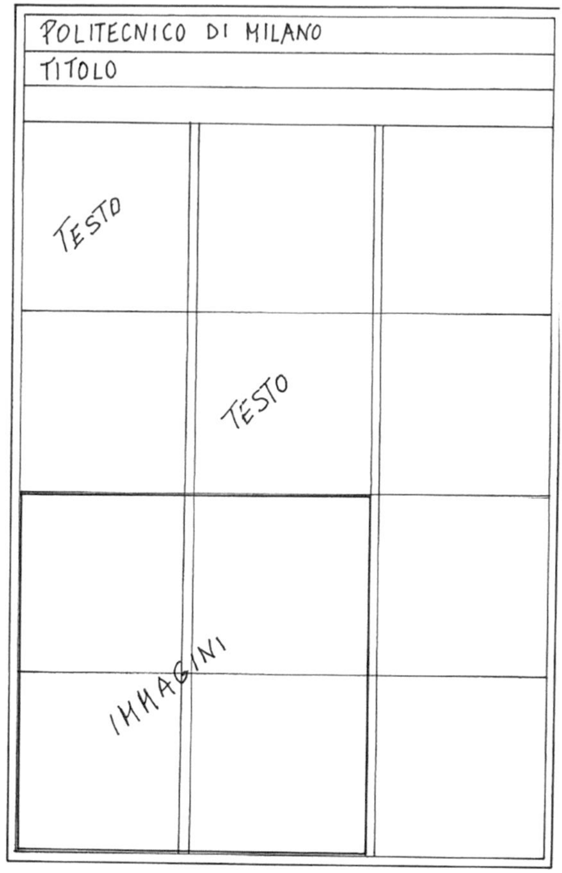

Fig. 4.2 Grid utilized for the exhibition for the 125th anniversary of the founding of the Politecnico di Milano

The words on the figure translate as: titolo (title), autori (authors), affiliazioni (institution), introduzione (introduction), metodi (methods), risultati (results), and bibliografia (bibliography).

In some cases, the pictures on the placard can be substituted with actual small objects. Such a placard, like the one utilized by the Author for teaching machine elements, can be more realistic, for the simple fact that an object represents a different level of reality than a picture.

Additionally, with a placard, it is possible to individuate the different paths of the museum, by means of:

- the language of the texts (Fig. 4.4);
- the interests of the users (Fig. 4.5);
- the level of the users (Fig. 4.6).

Another interesting configuration of placards can be placement of one placard with general information that is then supplemented by an auxiliary placard with

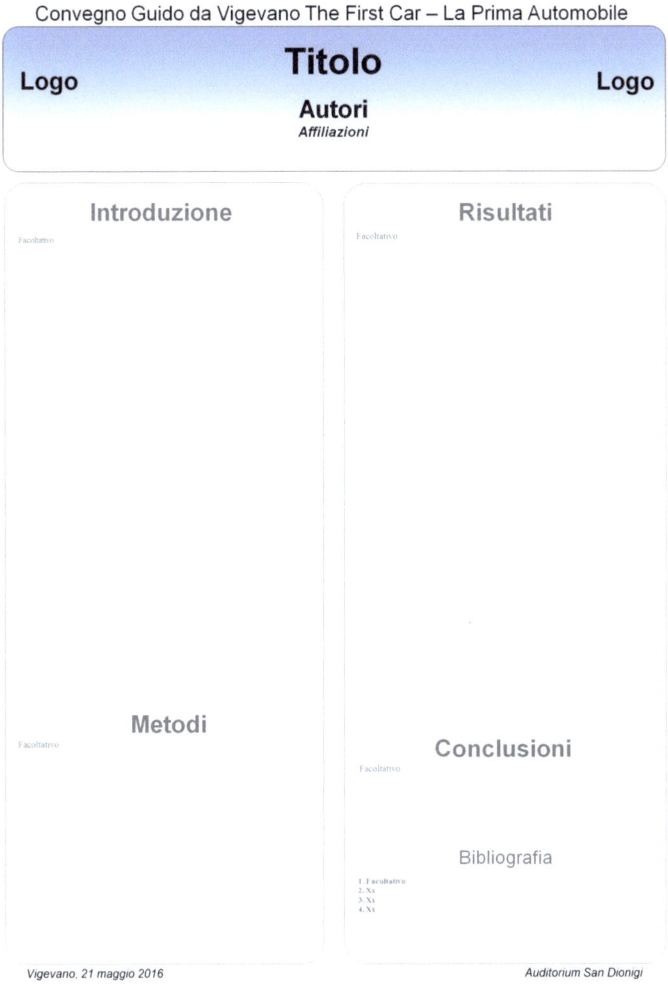

Fig. 4.3 Grid utilized for the exposition of a placard in a scientific congress

Fig. 4.4 Differentiation of paths on a placard based on language

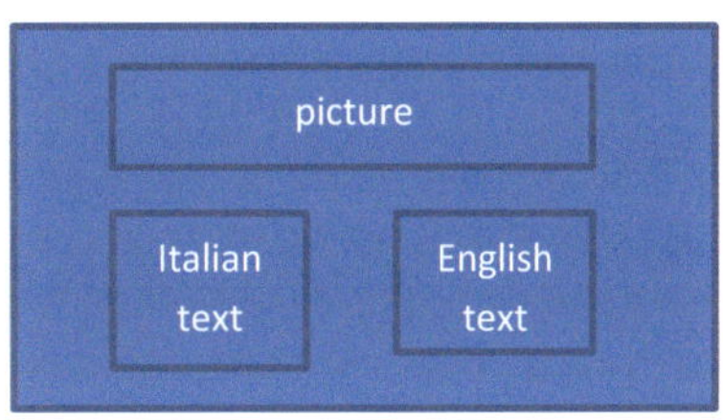

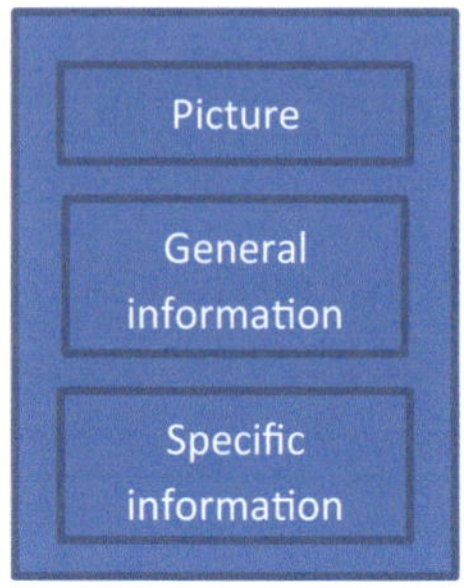

Fig. 4.5 Differentiation of paths on a placard based on general and specific texts, in relation to the different interests of the visitors

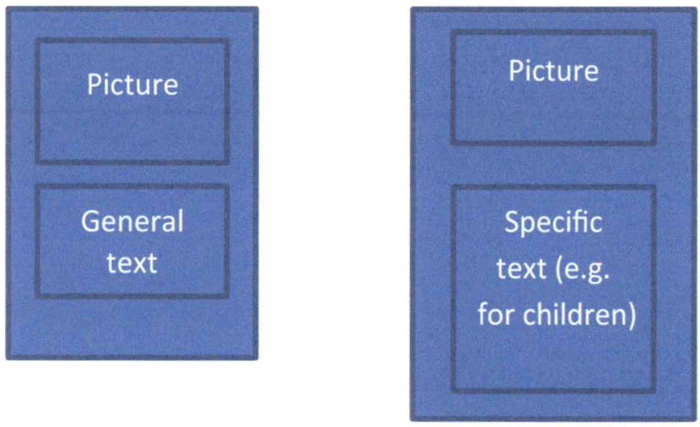

Fig. 4.6 Differentiation of paths on a placard based on a general text and a separate text for particular categories of visitor, e.g., children

more in-depth information. As an example, if one has a placard showing reproductions of old coins, it could be interesting to place another one nearby with guidelines on how to read the inscriptions on the coins.

4.3 Realization

An important step in the realization is the preparation of the first configuration of the placard, through use of the above-mentioned grids. Figure 4.7 is an example of the preparation, using the grid from Fig. 4.2, of a placard for the exhibition for the 125th anniversary of founding of the Politecnico di Milano. The specific focus of the placard is the First World War.

The different parts of a placard can be realized through the pasting of text and images (a solution used frequently in the past) or through the use of specific software, as is done most often nowadays. Such software tends to offer a general individuation of the relationship between text, pictures and blanks through use of templates of the varieties of grid presented above.

In Fig. 4.8, some options are presented.

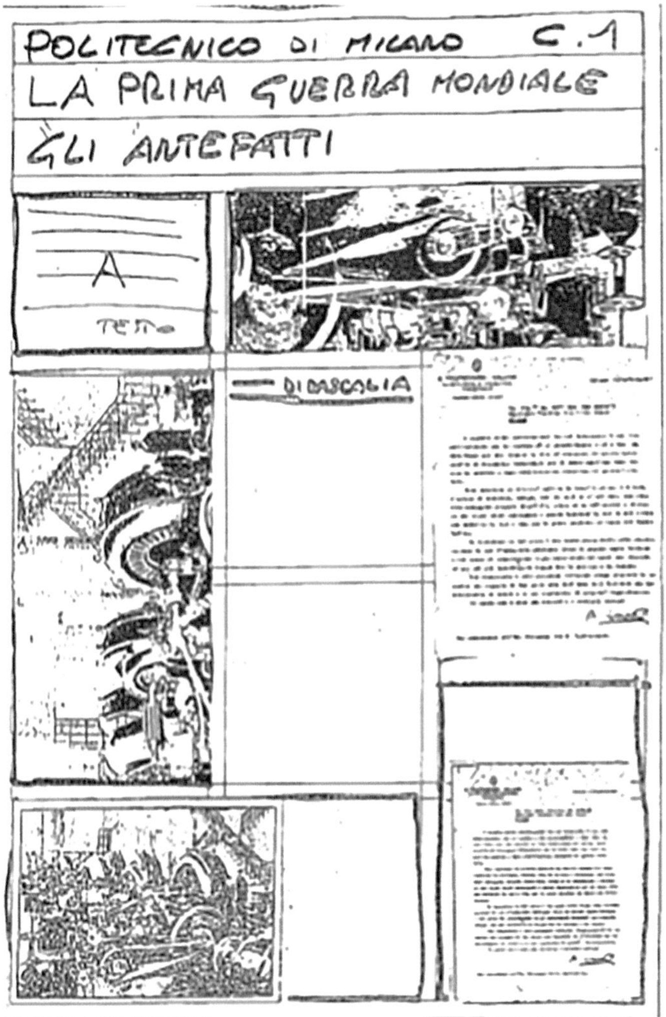

Fig. 4.7 Example of the preparation of a placard related to the First World War, using the grid from Fig. 4.2

The following are also very important considerations:

- the contrast between the font and the background colors must be very strong
- the text should not be too long nor too grammatically dense. For example, adverbs and adjectives tend to be generally unnecessary
- the characters must be congruent with the size of the readable range Table 4.1, as indicated in (Da Milano and Sciacchitano 2015).

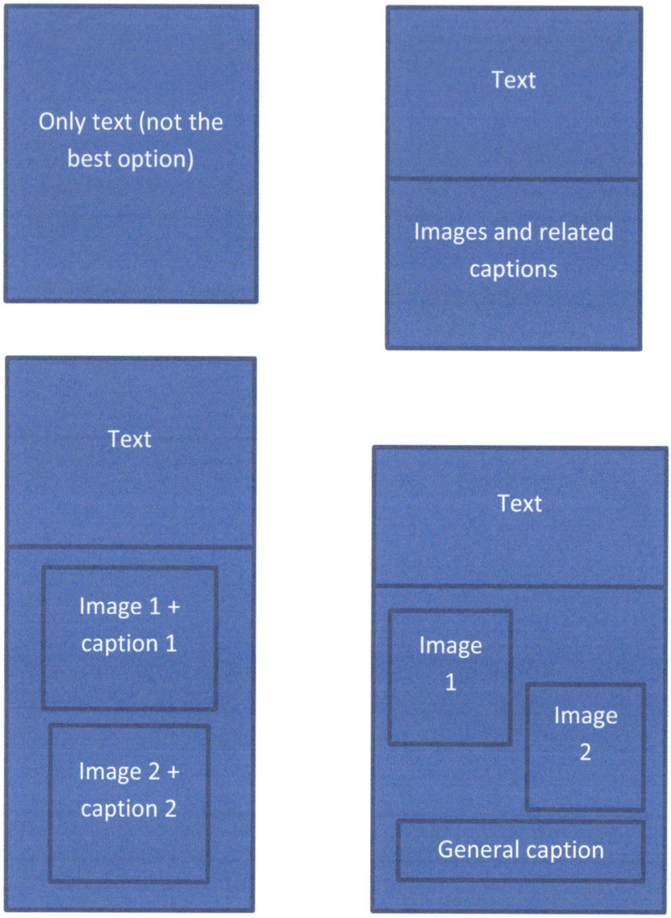

Fig. 4.8 Some examples of configurations of placards Texts must be carefully realized, so as to be congruent with the estimated initial behavior of the visitor, as should all other communication

Table 4.1 Size of characters related to the distance of the visitor

	Good lighting (mm)	Poor lighting (mm)	Very poor lighting (mm)
Visitor at 36 cm	0.76	1.3	2
Visitor at 71 cm	1.3	2.5	3.8
Visitor at 91 cm	1.3	3.3	4.8
Visitor at 105 cm	2.8	5.3	8.1
Visitor at 300 cm	5.6	11	16.5

- To minimize the effort required of the reader, it is advisable that the length of and number of words per sentence be reduced. One criterion for evaluating the difficulty of an English text is the formula by Flesch: the readability F of a text is given by the following relationship:

$$F = 206,835 - (0,864\ S) - (1,015\ P),$$

where S is the average number of syllables per word calculated in a sample of 100 words and P is the average number of words per sentence.

The general criterion is, of course, that a text be as easy to read as possible, using simple words and short sentences. The classification of the readability is as follows:

90–100: Very Easy
80–89: Easy
70–79: Fairly Easy
60–69: Standard
50–59: Fairly Difficult
30–49: Difficult
0–29: Very Confusing

- Start with the most important information.
- Avoid information overload.
- Stick to one type of character, using italics or bold to highlight key words only if necessary.
- Leave white space between lines of text.

In the reference (Da Milano and Sciacchitano 2015), it is possible to find more information about placards.

The most important utilization of the placard is, of course, the exposition in the museum and/or exhibitions. Another utilization can be reproduction of the placard for use within the catalog. As an example, the catalog from the above-mentioned exhibition for the 125th anniversary of the Politecnico di Milano was realized by utilizing the placards as pages of the catalog itself. In the next section, some further examples are presented.

4.4 Examples

In this section, some examples are presented and commented upon. These examples are placards from the exhibition for the 125th anniversary of the Politecnico di Milano and, as mentioned above, also pages from the catalog. The placards are very varied and represent an interesting sample of the possibilities for configuring a placard.

Figure 4.9 is related to the general configuration of the engineering courses. The rule is implicit in the title and the example is a schema of courses and relative subdivisions.

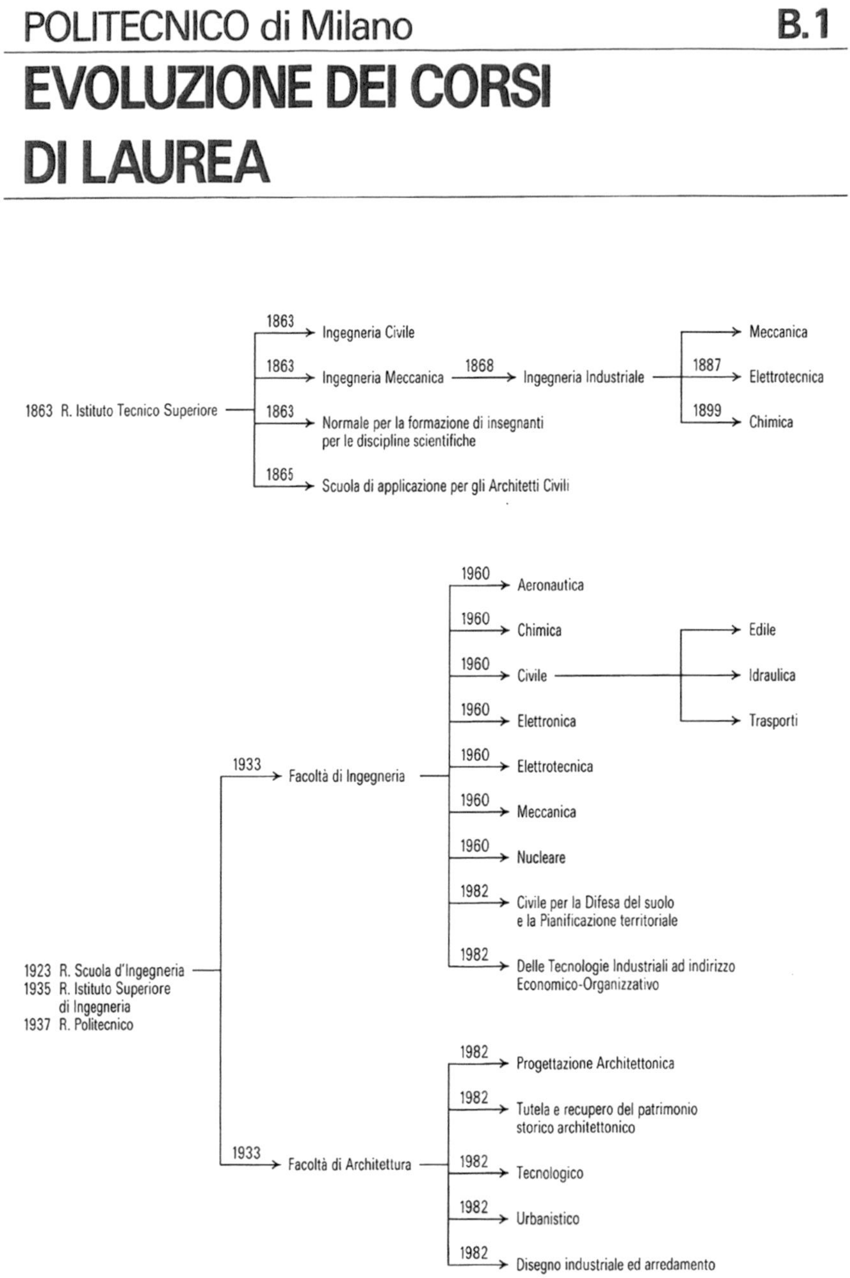

Fig. 4.9 Placard related to engineering courses

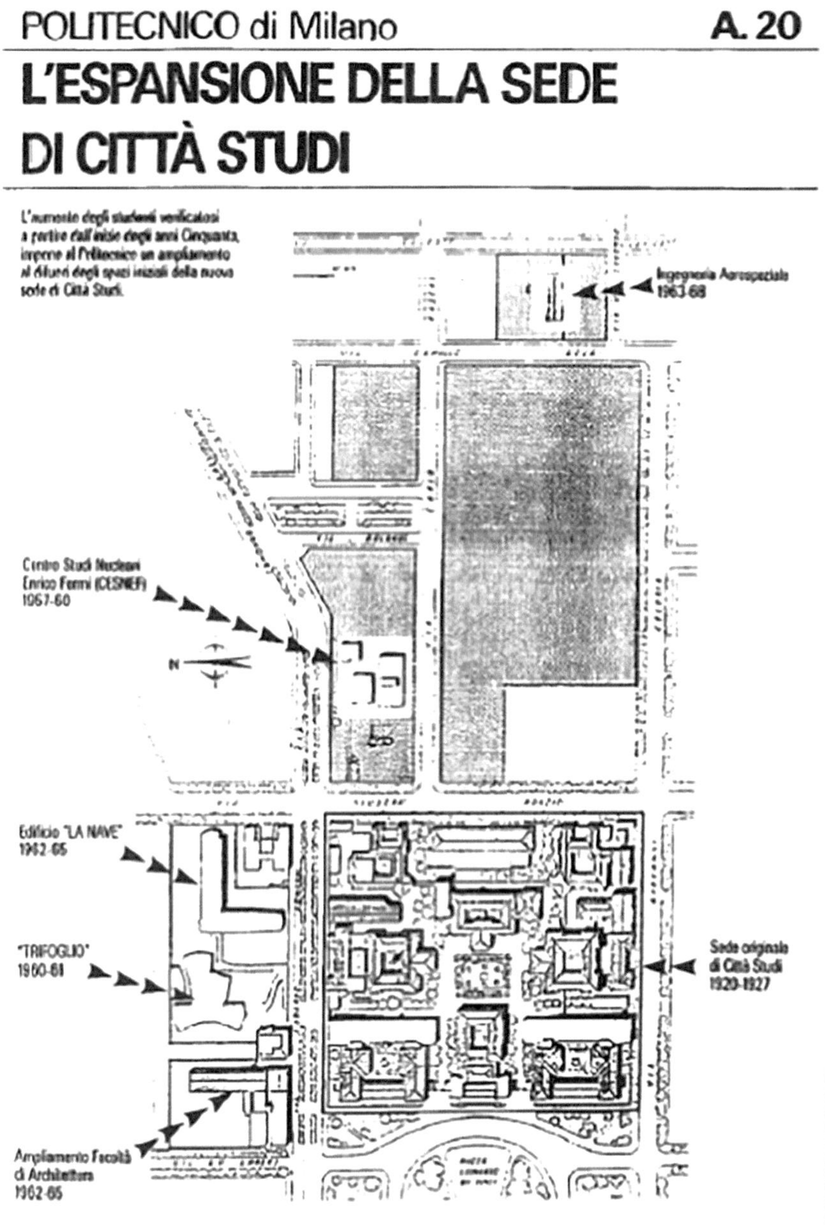

Fig. 4.10 Placard related to the initial location of the Politecnico di Milano

The placard in Fig. 4.10, related to the initial location of the Politecnico di Milano, is very similar in general configuration to the placard in the previous figure. The rule is very short and the example is constituted by a map of the neighborhood

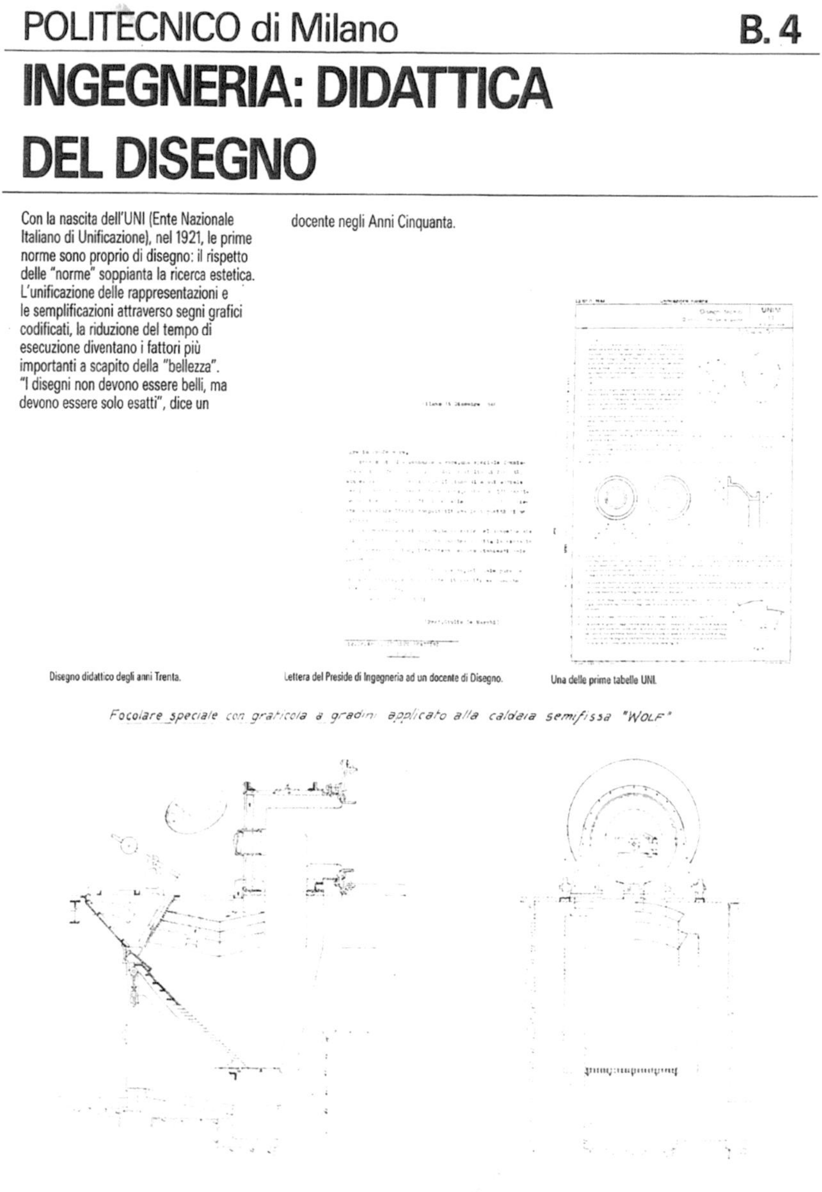

Fig. 4.11 Placard related to the teaching of drawing

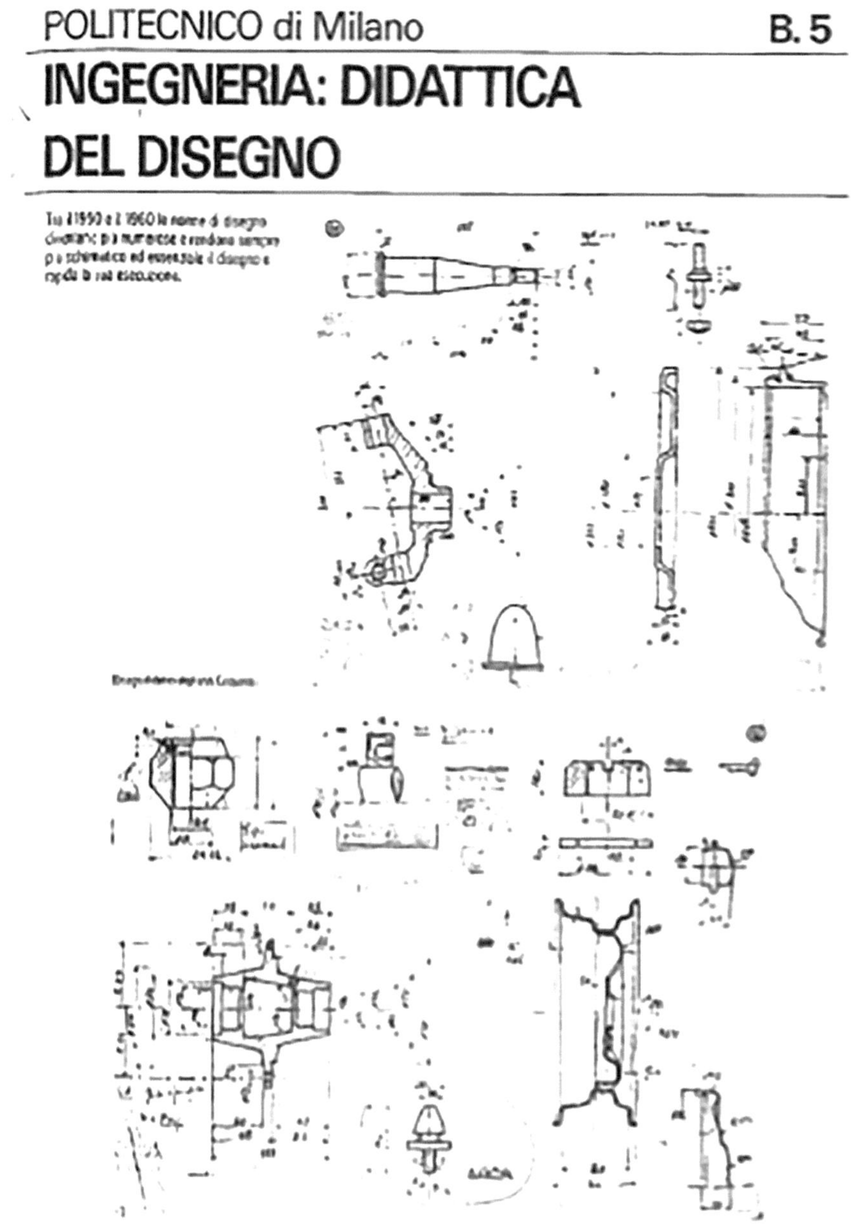

Fig. 4.12 Placard related to the teaching of drawing

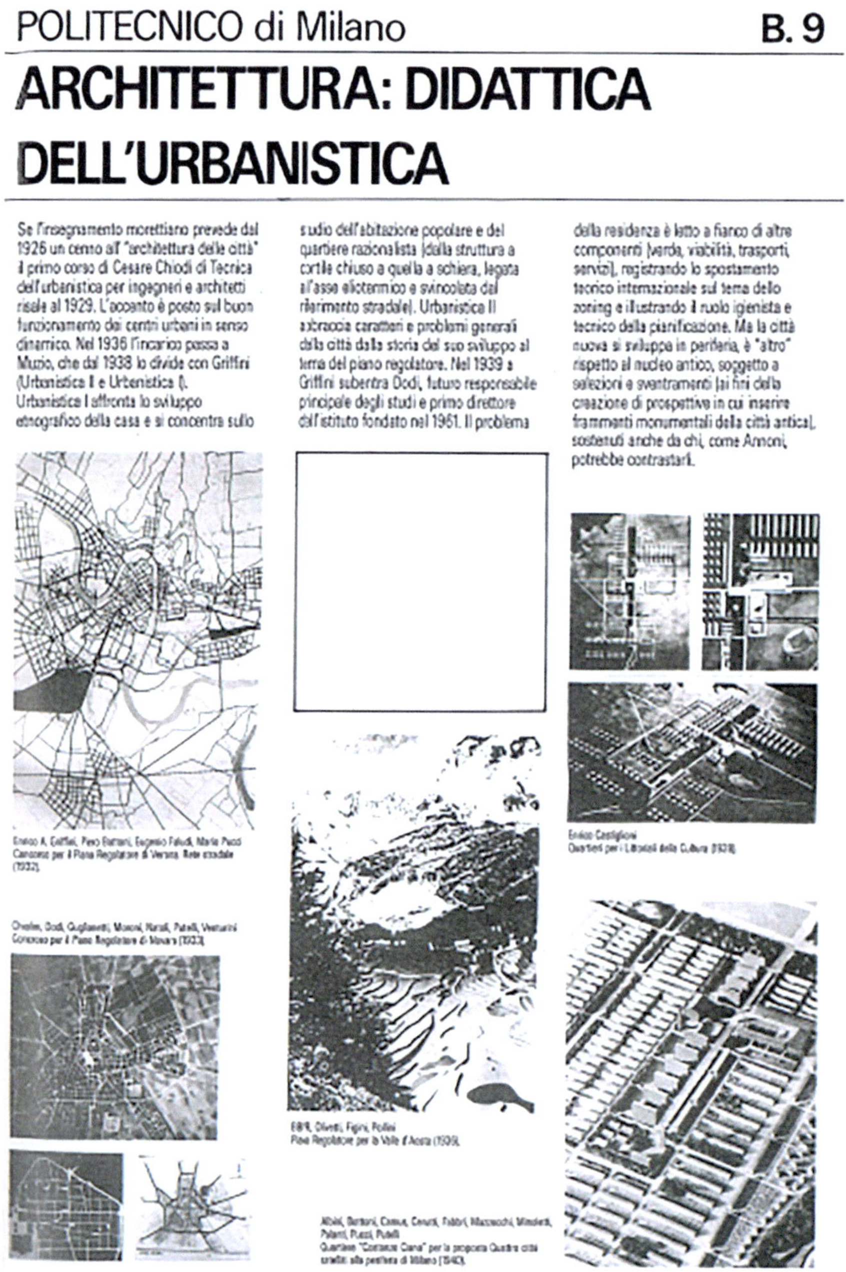

Fig. 4.13 Placard related to the teaching of city planning

POLITECNICO di Milano D.15

TECNOMASIO ITALIANO BROWN BOVERI

Il Tecnomasio fondato, come il Politecnico, nel 1863 e divenuto nel 1903 Tecnomasio Italiano Brown Boveri, è la più antica Società elettromeccanica italiana ed una impresa leader nel settore delle macchine elettriche, degli impianti elettrici, dei veicoli ferrotranviari.
I legami del Tecnomasio con il Politecnico non si limitano alla stessa data di nascita, né alla realizzazione dei primi strumenti di fisica ed elettrologia: la Società ha assorbito molti laureati del Politecnico che vi portano il loro contributo di cultura, di intelligenza, di operosità. Alcuni di essi, poi, seguono gli allievi nelle esercitazioni e nella preparazione della tesi di laurea.

Mario Giuseppe Soldini (1894-1968), laureatosi al Politecnico nel 1918, nel 1952 per volontà di Giacinto Motta, entra alla STEWAG, in Austria, occupandosi dell'esercizio di impianti elettrici. Nel 1938 passa al Tecnomasio, dove contribuisce in modo determinante sia a superare il difficile periodo della guerra, sia ad affrontare i problemi della ricostruzione.

Virginio Gandini, nato a Milano nel 1904 e laureatosi al Politecnico di Milano nel 1927, è direttore centrale del Tecnomasio Italiano Brown Boveri. Nel 1954 è socio promotore e fondatore dell'ALDAI (Associazione Lombarda Dirigenti Aziende Industriali). È Presidente dell'EMIT (Ente Morale Feltrinelli per l'incremento della istruzione tecnica), ed opera in collaborazione con il Politecnico. Medaglia d'oro al merito direttivo e Maestro del Lavoro, riceve la medaglia d'oro di benemerenza dal Ministero della Pubblica Istruzione per la scuola, la cultura e l'arte.

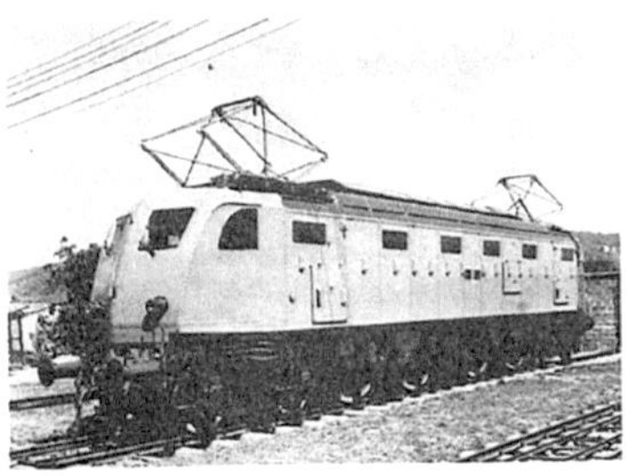

Fig. 4.14 Placard related to an important Italian factory that was active in the construction of railway vehicles

Fig. 4.15 Placard related to the connection between the Politecnico and the Pirelli rubber company

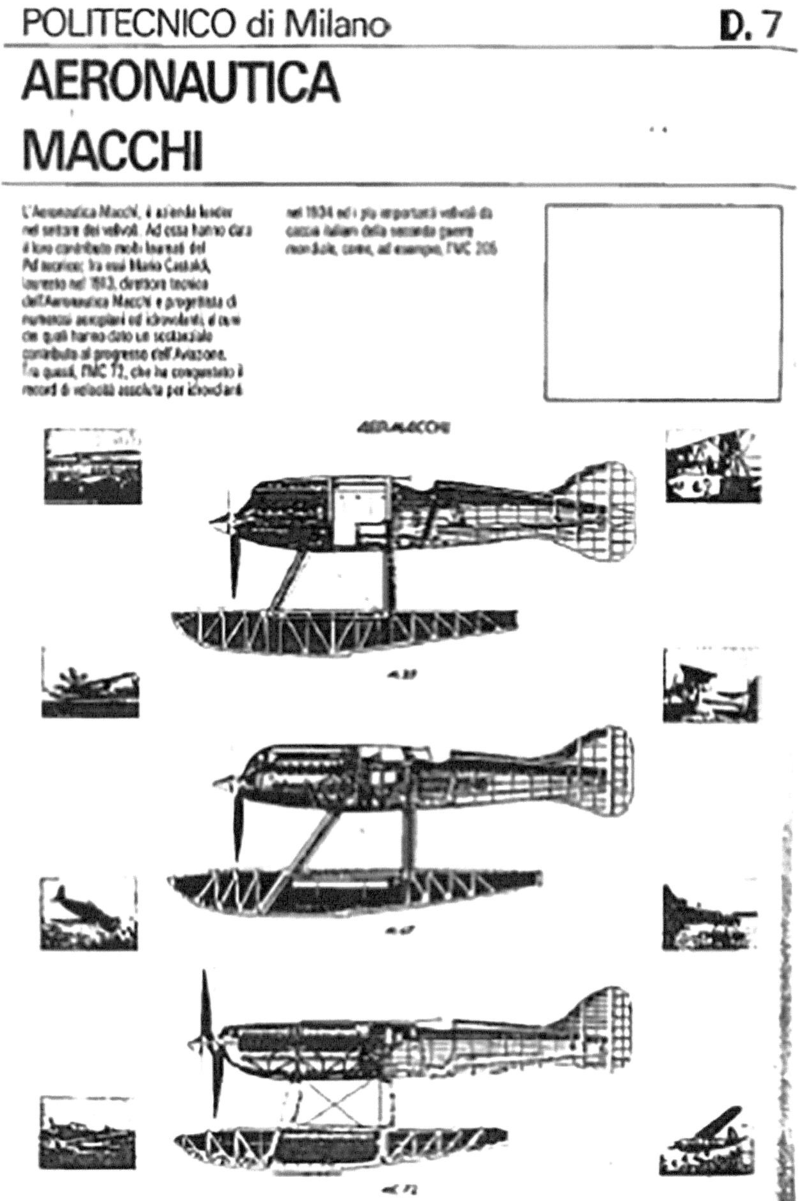

Fig. 4.16 Placard related to the company Aeronautica Macchi

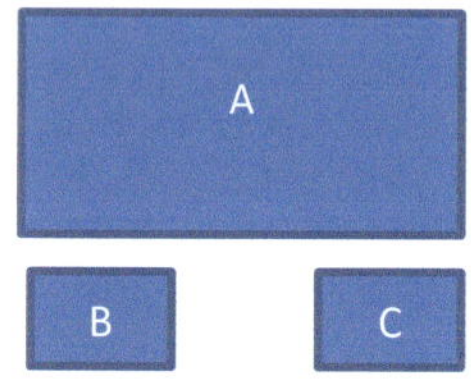

Fig. 4.17 Example of a placard characterized by a big picture in A and two videos in B and in C

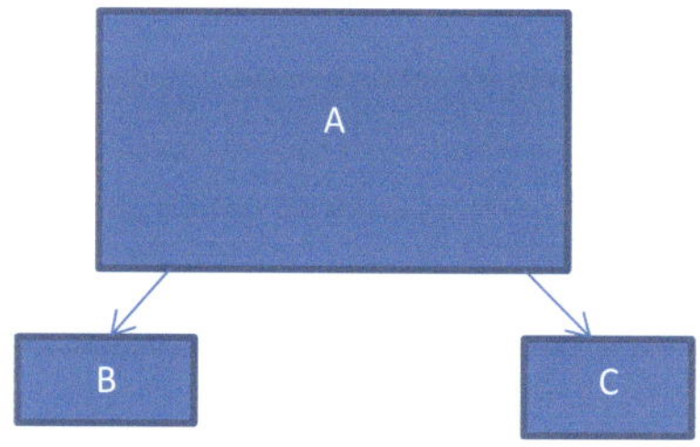

Fig. 4.18 Example of a placard with information constituted by images and text (A), supplemented by two enlarged details (B and C)

of Milan where the Politecnico was founded. On the map, the locations of the buildings of the Politecnico are highlighted.

Figures 4.11 and 4.12 are related to the teaching of mechanical drawing at the Politecnico di Milano. The rule is a general description of the fundamental steps in the teaching of drawing, and the examples are actual drawings by students.

Figure 4.13 shows a placard devoted to the teaching of city planning at the Architectural Faculty. It is important to observe that the rule is much longer (perhaps too much!) than that in the previous placard.

Figure 4.14 is related to an important Italian factory, active in the construction of railway vehicles. The rule is constituted by some biographical information. The example is constituted by portraits of the two most important engineers at the factory, both of whom graduated from the Politecnico. Photos of certain significant products have been placed adjacent to the portraits.

Figure 4.15 shows a placard detailing the links between the Politecnico and the well-known Italian rubber company Pirelli, which was founded by an "old student" of the Politecnico, Giovanni Battista Pirelli.

The rule, in this example, is made up of text, while the examples are significant documents related to the activity of the company and an advertising poster.

Another important "old student" of the Politecnico was Mario Castoldi, manager of Aeronautica Macchi and well-known designer of airplanes and seaplanes. Figure 4.16 depicts a placard devoted to Mario Castoldi. The rule is constituted by some biographical information, while the example is constituted by pictures of Castoldi's most important creations.

Figures 4.17 and 4.18 show some further examples of the configuration of placards.

Reference

Da Milano C, Sciacchitano E (2015) Linee guida per la comunicazione nei musei: segnaletica interna, didascalie, pannelli Quaderni della valorizzazione MiBACT (Ministero dei Beni e delle Attività Culturali e del Turismo), Roma

Chapter 5
Presentation of Objects

5.1 General Principles

The objects are the fundamental part of a museum and their related collections. Therefore, their importance requires careful presentation and explication. The objects should, in general, be exhibited in a proper environment, so as to transmit precise information and sensations to the visitor. The visitor, through observation and consequent critical elaboration, should experience the sensation of being absorbed into the reality represented by the object, e.g., to feel what is was to live in its time period and in the working environment in which it was created.

Some general exposition criteria are as follows:

(a) It is important to establish the correct *density* of the exposition, congruent with the ease of observation. Too great a number of objects in a limited space will hamper proper observation. The most significant objects must be exhibited in isolation, so as to highlight their importance and to concentrate the attention of the visitor upon them.
(b) It is also useful to locate the different objects on different planes and along different lines of sight, so as to compel the visitor to continuously vary their eye focus.
(c) The height of the exposition should be between 1.00 and 1.80 m: upper and lower heights are not congruent with good observation.
(d) The showcases should not have interior lighting, because certain sources of light can, in many cases, damage certain objects. In relation to the nature of the protected objects, the inside of the showcases must constitute a microclimate, with controlled temperature, humidity and dust.
(e) If the objects require repair, historical objectivity must be privileged over aesthetics.
(f) Sometimes, visitors may attempt to get too close to the objects; in such cases, it is necessary to protect the objects. If it is impossible to shield the objects within

A. Rovetta and E. Rovida, *Scientific Knowledge Communication in Museums*,
Research for Development, https://doi.org/10.1007/978-3-319-68330-0_5

a showcase, some alternate forms of protection can be as follows: platforms, ropes, stairs, transparent walls.

5.2 Dimensions

5.2.1 *One-dimensional Objects*

Examples of such objects are swords, spears, cables, wires, and bars. They can be exposed vertically, or hung from the ceiling, or laid out horizontally on proper supports.

5.2.2 *Two-dimensional Objects*

Examples of such objects are textiles, paintings, fabrics, and drawings. They can be exhibited in vertical or horizontal arrangements, in relation to their size and to the type of object. If they are not sensitive to light, they can be exhibited in showcases that are either stationary or that rotate, like the pages of a book.

If, however, they are sensitive to the light, they can be located in vertical or horizontal chests of drawers to be opened according to the needs of observation.

5.2.3 *Three-dimensional Objects*

Such objects represent the majority of objects of interest from the scientific-technical point of view.

5.3 Size

Objects can be classified in regard to size, as indicated in Table 5.1.

Table 5.1 Classification of objects according to size

Class of object	Dimensions (m^3)
Very small	Smaller than 10^{-6}
Small	10^{-6} until 0.0085
Medium	0.0085–0.03
Large	0.003–0.3
Very large	More than 0.03

5.3.1 *Very Small Objects*

The fundamental problems regarding this class of object relate to defense from damage and the diminished possibility for observation. The objects must be protected in showcases equipped with burglary systems and alarms. Another possibility is to place the objects in a cavity formed between two transparent layers.

Visibility can be facilitated through magnifiers or videos that allow the visitor to explore the details of the exhibited object. An even simpler system can involve use of one or more photo enlargements of the most significant details.

5.3.2 *Small Objects*

Such objects present the fundamental problem of protection against theft; there are, in general, no problems in regard to observation.

Small and very small objects can be placed within crates, openable by the visitor and, of course, equipped with safeguards.

5.3.3 *Medium and Large Objects*

Considerations for such objects are generally analogous to those for the small ones.

5.3.4 *Very Big Objects*

The problem in the case of this class of objects can be the perception of synthesis, which, with such great sizes, can be difficult. The problem is the realization of an adequate path around or inside the object.

One solution can be the realization of a reduced physical model or a picture, or a virtual model of the original object.

Another solution could be the realization of a model corresponding to half or a quarter of the object, which, through a system of mirrors, gives the impression of the entire object.

5.4 Requirements of Observation

The requirements of observation can be very different in relation to the nature and typology of the objects.

5.4.1 *Objects to Be Observed from Only One Side*

In this case, in general, there are no problems.

5.4.2 *Objects to Be Observed from Both Sides*

This problem can be solved through use of a mirror, which can be fixed or actuated by the visitor. As an alternative, showcases with mirrored glass could be used. In other cases, it is possible to utilize rotary platforms on which the object is placed. Further alternatives can be to use photos or video systems to show the otherwise invisible side.

Another solution could be the presentation of two identical objects, presented with opposite orientation.

5.4.3 *Objects to Be Observed Internally*

Many objects have internal spaces that are interesting to observe. If there are no natural openings present through which observation can be effected, the objects can be presented in a sectioned state or side by side with a sectioned model. Otherwise, there is always the possibility of presenting photos or videos displaying the internal details alongside the original object.

If the objects are very large (e.g., vehicles or ships), realization of an internal path becomes that much easier.

5.4.4 *Objects in Their Environment*

Objects presented in their natural/logical environment are very interesting from the points of view of documentation and education. Such objects, in effect, are presented in a context that highlights scenery and connections in realistic and comprehensive situations. In this way, the presentation of the objects becomes more effective. Such presentations, in addition, can often deliver greatly enhanced emotional impact.

A classic example is the diorama: these are scenographic, three-dimensional reconstructions of natural landscapes, working environments, or historical situations in which the presented objects would most likely be found. For example, in an archeological museum, an exposition of funerary objects (such as vases, weapons, and artifacts) could be positioned within a reconstruction of an ancient tomb.

The elements of a diorama, in general, are as follows:

(a) Objects (models or real objects);
(b) Environment (e.g., workshop, factory with concrete elements that highlight the actual functioning of the presented objects);
(c) Background (the full environment);
(d) Explanation placard (or video), with some or all of the following information:
  (d.1) name of the object/s;
  (d.2) time of realization;
  (d.3) location of realization;
  (d.4) name of the inventor/manufacturer;
  (d.5) some information about the construction and configuration of the object;
  (d.6) some information about the function of the object;
  (d.7) some information about the object's historical role.

Dioramas can be fixed or feature movement. In the second case, which is more interesting from the documentary point of view, the presented objects (e.g., machines or mechanisms) move according to working cycles that correspond to the real situation. Such presentations, often the result of cooperation between scientists and artists, are instructive in a pleasant way. They can also be complemented with placards with texts, photos, videos or multimedia systems. In some cases, spoken commentary is included, either recorded or performed live by an actor, perhaps dressed in the fashion of the period of the presented object. The spoken commentary can be accompanied by directed light, so as to highlight the particulars of the diorama in conjunction with the spoken commentary. A musical accompaniment or a system of sounds associated with the exhibited objects can also be useful, particularly from the emotional point of view.

The movement within the dioramas can be continuous or activated by the visitor.

Analogous to the dioramas in certain respects, there are also plastic reproductions, on a small scale, of landscapes and environments. These reproductions can be animated by moving vehicles, such as trains. There is also the possibility that individual sections can be highlighted by the visitors, by, for example, activating available lights.

Dioramas and small-scale reproductions, with their emphasis on detail, can be a source of great emotional involvement for the visitor.

5.5 Requirements of Interactivity

The relationship between the visitor and the exhibited object can be very different now than it has been in the past. For many years, a museum was only a collection of *passive* objects; the visitor was only a passive observer. Nowadays, both roles have undergone a profound change, and the visitor is very often encouraged to interact with the exhibited objects. According to the requirements of interactivity, exhibited objects can be classified as follows.

5.5.1 Objects Only There to Be Seen

In this case, there are no requirements of interactivity and, consequently, from this point of view, no problem.

5.5.2 Objects There to Be Seen and Touched

In this case, the requirements of interactivity are very low and, as long as there is no danger of damage to the objects, there are no problems.

5.5.3 Objects There to Be Operated by the Visitor

A widespread trend in museums today is the possibility of the visitor interacting with the exhibited objects. In this way, the old philosophy of a *static* museum becomes outdated and is progressively replaced by the *dynamic* museum, where the visitor can interact with the object in a genuinely meaningful way.

The possibilities for interaction can be varied: e.g., the visitor can activate a machine or a mechanism, perform an experiment, or *enter* into a technical situation through virtual or augmented reality.

In general, it is possible to distinguish certain specific types of case:

(a) The exhibited object is activated directly by the visitor by means of a specific part, such as a lever, crank, or wheel. In this way, it is possible, for example, to give a hands-on lesson on the mechanics of the machine or mechanism. In some cases, the possibility of interaction with a real object is virtually uninhibited, e.g., children play with models of weapons from the "Star Wars" movies.
(b) The exhibited object is activated by an electric motor, initiated by the visitor through the push of a button.
(c) In place of the object, there is a multimedia system upon which it is possible to perform physical experiments.
(d) In place of the object, there is a virtual model, through which it is possible to perform experiments and simulations.

5.5.4 Objects There to Be Used

In such cases, the observation is very complete, because it is possible to see the object engaged in its actual operation, or at least a simulation thereof. What is now referred to as a "living museum" is predicated on such situations in which certain activities are demonstrated through the use of instruments or machines in their normal function.

The possibility of the visitor carrying out certain direct experiments can be very important, particularly from the emotional point of view. For example, the experience of the visitor in a flight simulator or in a (soft!) crash test, or climbing into the cockpit of a locomotive, or sending a telegraphic message with Morse code, can be unforgettable.

5.5.5 *Objects There to Perform a Given Experiments or Application*

In these cases, it is possible to see, at certain defined times, experiments conducted by the staff of the museum with corresponding explanations.

5.6 Features of an Exhibiition

5.6.1 *Objects Exhibited*

This is, of course, the normal situation.

5.6.2 *Objects in Storage*

Such objects could be considered for future expositions if specific investigation indicates that they would be considered interesting by the visitors: if such interest is high, the above-mentioned objects could be considered for exhibition during the phase of redesign of the museum.

5.6.3 *Objects Only Exhibited on Specific Occasions*

Some objects in storage could be exhibited only on particular occasions, such as specific exhibitions related to the nature and/or features of the objects themselves.

5.6.4 *Objects not Exhibited Because They Are in Restoration or Have Been Exhibited Elsewhere*

Such objects can be represented by photos or virtual presentations, showing, e.g., phases of the restoration or the transport of the above-mentioned objects to the temporary exhibition to which they have been loaned.

5.6.5 *Objects of Particular Importance*

Objects of particular importance can be presented supplemented by placards featuring material such as:

(a) historical/biographic information: author, time, how the object was created (cultural climate, similar experiences, progress represented)
(b) general description: description of the objects
(c) particular description: highlighted details, with deeper descriptions of each of them
(d) new general description: a more deep general description, with reference to the above-presented details.

In addition, the object itself can be enhanced with the inclusion of sounds and lights, with the aim of (re)creating an emotional environment that will grab the visitor's attention and stick in their memory.

5.7 Presentation of a Particular Object

5.7.1 *General Presentation of an Object*

A proposal for the general presentation of an object can be found in Fig. 5.1.

The visitor, finding themselves in front of the object, can follow three different routes:

(a) Direct observation of the object, with captions and a placard with general explanations. If the object is complex, it can be useful to include a simplified small-scale model to highlight the configuration and functioning of the actual object. Obviously, the captions and placards may be placed in slightly different areas, and therefore present different paths for the visitor to follow as they read, depending on their individual needs and interests.
(b) Virtual interaction with the object, with the possibility of watching it set in motion, as well as seeing it disassembled and reassembled and interacting with the different pieces.

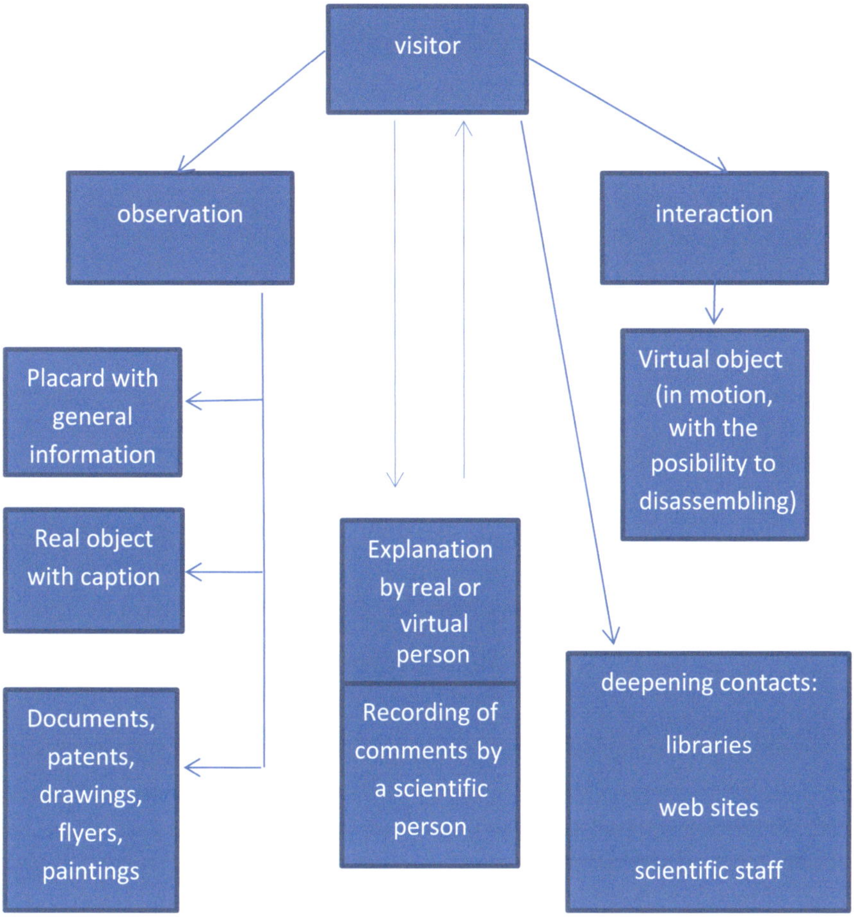

Fig. 5.1 General presentation of an object

(c) Opportunities for deeper investigation, through links with libraries, websites, contact with the scientific staff of the museum, etc.

5.7.2 *Books*

Books can be exhibited as follows:

- closed, so as to show the cover and the binding
- open, so as to show the two most interesting pages

- open as indicated in (b), but with photos of other interesting pages located nearby.

The captions accompanying the books should include the following information:

- title
- author/s
- editor
- place and year of publication
- library of origin/cataloging/marking
- dimensions
- historical events/significant owners
- historical importance
- scientific importance
- bibliography
- in-depth bibliography
- notes.

Particular care is, of course, necessary to protect the book from light, dust and atmospheric pollution. Also very important is the inclination of the book in relation to the direction of observation.

5.7.3 Documents

Documents should be exhibited in showcases and carefully protected from light, dust and atmospheric pollution. The original can be accompanied by a photo of the side hidden from the viewers. Another solution can be the exhibition of the document in a vertical arrangement, with both sides visible.

If the document is constituted by many pages, the most important side of the original can be exhibited, while the other most important pages can be presented through photos.

If the document is not easily readable (for example, old documents in gothic characters), it can be useful to include a translation in the language of the country where the museum is located (it would also be standard these days to include one in English as well).

5.7.4 Vehicles

Vehicles are particularly suited to being presented in an appropriate environment. The carriage, e.g., can be exhibited with mannequins of horses, riders and coachmen. A reconstruction of an ancient stable could be a very interesting setting.

Cars can be presented on a reconstruction of a road. A mirror placed under the car can allow for the undercarriage of the car to be observed.

The presentation can be completed with a placard featuring technical data, original photos, flyers, etc. Next to the cars, an exposition of automotive components, such as tires, radiators, and carburetors, can add further interest.

A useful setting for an exhibition of cars could be a reconstruction of a service station or garage.

Railway cars can be exhibited on a section of tracks, or in an abandoned station (or a reconstruction thereof). Stairs and walkways set up at the side of the railway car can add further interest by allowing the visitors to look inside.

Airplanes can be exhibited on tarmac, as at an airport, or suspended from the ceiling by means of rods.

5.7.5 Ships

Ships of small dimension could be presented in a water basin. Full-size ships, however, would need to be presented outdoors, either beached or in a large shed, accompanied by an exposition of placards, photos, and virtual models depicting significant details.

5.7.6 Instruments

Instruments are objects of small or medium dimensions, utilized with technical expertise in industries, laboratories, and workshops. The presentation of such objects can be complemented with placards giving explanations of their constructive and functional details. Technical drawings can also be useful for virtual reconstruction of their utilization.

5.7.7 Reconstructions of Work Environments

Reconstructions of work environments (laboratories, factories, mines, industries, firms, offices, housing) are constituted by various objects grouped together organically. Such reconstructions are very difficult to be realized, because they require a number of complete collections; in addition, it is necessary to have thorough documentation of the considered field and period of time, with the aim of guaranteeing the fidelity of the reconstruction.

The above-mentioned reconstructions are interesting because they are a documentation of the life and activities from their given period; in addition, they allow visitors to experience a simulation of a certain environment, stimulating not only cultural, but also psychological and emotional involvement.

Such reconstructions can be, in general:

- an actual realization, such as, e.g., disused factories or mines;
- reconstructions on a scale of 1:1;
- scale models.

It is very important, in the reconstruction of work environments, to highlight the emotional aspects, so that the visitor may get a sense of the psychological involvement of the people who were active in such an environment. As an example, at the chimney sweep museum in Santa Maria Maggiore, a mountain village in Northern Piedmont, the presentation of sounds and songs from the life of the chimney sweeps (very often children destined for a hard life) add a true element of poignancy and leave an indelible imprint on the visitor.

Other examples could be a museum dedicated to a school, showing the typical daily movements of the students, creating atmosphere through such touches as ringing the bell for start of lessons, a portion of a lesson being presented, etc., or a museum dedicated to transport, with the exhibits being embellished by traffic noise, horns, sirens, etc.

5.7.8 Evolution of Historical Revelations/Constructive Solutions

The evolution of a historical revelation and/or constructive solution can be defined as an ordered collection of the systematic development of a machine or mechanism. Such collections are interesting from two points of view: the first one is that they allow us to view the evolution of the constructive technique in chronological terms, while the second one is the possibility of comparing different constructive solutions and, consequently, developing interesting input for the creation and ideation of further innovative solutions. One problem in the realization of such an exhibit is the systematic determination of all significant solutions, ensuring that nothing relevant is omitted and, at the same time, that nothing that seems relevant, but is, in fact, irrelevant, is included.

Table 5.2 shows an example involving the historical evolution of car suspensions (Biggioggero et al. 2003).

It could also be interesting to show the historical evolution of a constructive solution or industry through a chronological depiction of the relevant advertising of the time.

Further possibilities for presenting the historical evolution of a constructive solution or industry (or a town, a people, etc.) are as follows:

- continuous sheet with pictures and texts in chronological order, representing a timeline

Table 5.2 Example of the historical evolution of car suspension

Year	Solution	Comment
1920		Concept by Vincenzo Lancia
1922	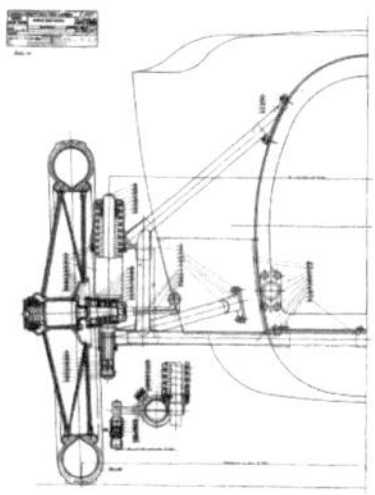	Lancia Lambda
1927	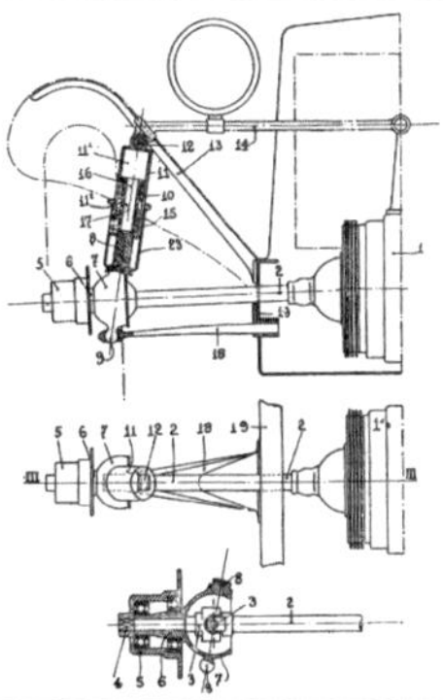	Fiat patent
1932	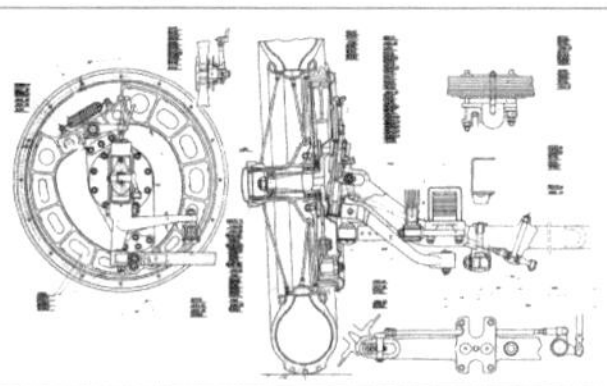	Alfa Romeo 8C 2300
1940	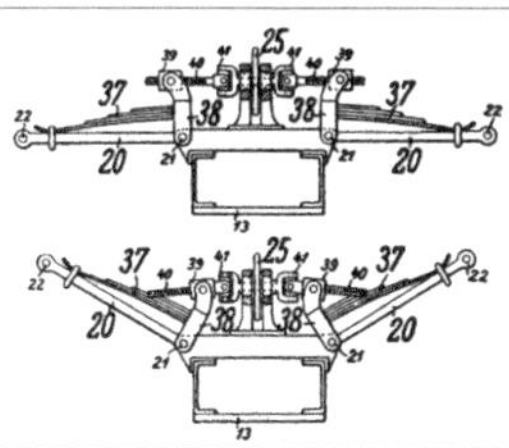	German patent

(continued)

Table 5.2 (continued)

Year	Solution	Comment
1961	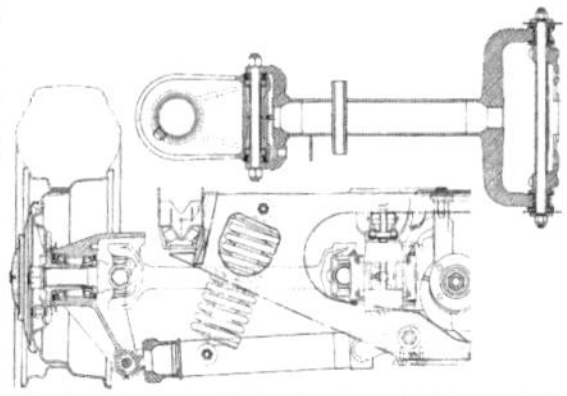	Jaguar MKX
1991	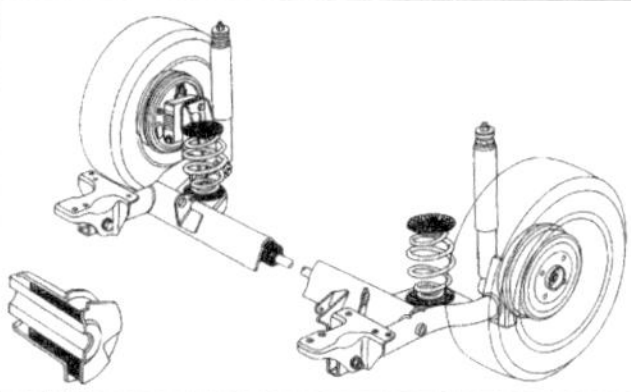	Renault Clio II
1995	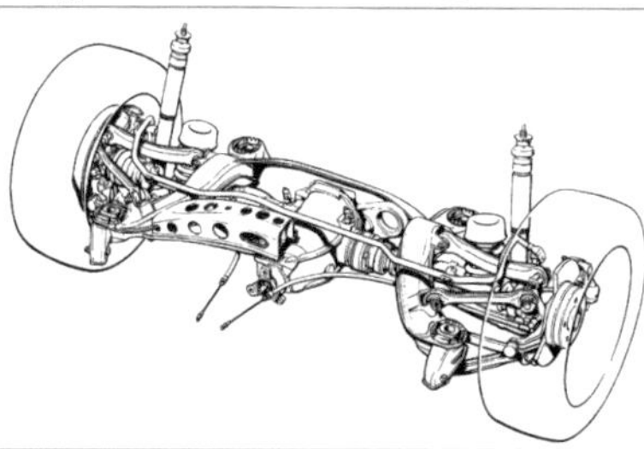	Mercedes Benz Classe E

- enlarged pictures of the most interesting events
- showcase with objects (and/or virtual models), accompanied, of course, by captions related to the above-mentioned events.

5.7.9 Fabrics

Fabrics can present in a great variety of products: lace, tapestries, embroideries, carpets, banners, full gowns. The fundamental problem is the delicacy of such objects, which require great care to protect them against light, humidity, heat and atmospheric pollution.

The raw fabrics can be exhibited in a horizontal disposition, e.g., in chests of drawers, or in a vertical disposition, e.g., suspended. Full gowns, however, can be exhibited in showcases on mannequins, hangers, or on an inclined plane.

Very large swaths of fabric can be presented hanging on a wall or from the ceiling. Reproductions of fabrics can be presented horizontally on the floor like carpets.

Semi-finished fabrics can be presented in big bags.

5.7.10 Surveys About a Period of Time

The reconstruction of different aspects of a period of time can be constituted, for example, as follows:

- Houses
- Offices
- Bars
- Leisure activities
- Cars and motorcycles
- Rail vehicles
- Ships
- Interior architecture
- External architecture
- Art
- Advertising
- Current events, news
- Furniture
- Music
- Telecommunications
- Lighting
- Appliances
- Cinema
- Fashion
- Important/significant persons.

It can be useful to present the above-mentioned aspects as a timeline, relative to the considered period of time.

5.7.11 Lives of People

An exhibition or a section of a museum devoted to illustration of the life of an important person should include, in general, the following information:

(a) An introduction with the aim of presenting the cultural, scientific, political, and historical climate of the period of time corresponding to the life of the person; such an introduction could include, e.g., documents, books, objects, texts and any other type of witness to the period of time. The introduction could be anchored by a characteristic image (e.g., a portrait or bust of the person) and/or reproduction of a particular quote or idea associated with the person.

(b) Essential biographical data, such as chronology, essential activity, particular accomplishments personal facts, geographical map indicating the significant trips the person took, etc.

(c) Detailed information about activities (scientific research, industry, production) and accomplishments (with emphasis on their role and significance in relation both to the specific time and successive developments).
(d) Papers, books and texts written by the person, with emphasis on the originality they display for the time and their importance in successive developments.
(e) The individual's more human qualities, including their relationships with other members of the scientific/technical/cultural world, their national and international engagement, their personalities in regard to culture and ethics, etc.
(f) Movies relevant to the person in regard to significant moments in their life or even simply to the historical location/era in which they lived.
(g) The involvement, whenever possible, of a living person somehow linked with the person to whom the exhibition (museum) is devoted, allowing for the presentation of direct witnesses, memories, or connections with the subject.

It is always a major boon for such museums to be able to be located in a building where the person lived. In such cases, the objects collectively tell the story of the life of the person: furniture, clothes, paintings, family documents (e.g., letters, photos, certificates), documents related to projects and activities, etc. As an example, the museum devoted to the Italian writer and poet Alessandro Manzoni is located in the writer's ancient house in Milan. Similarly, the museum devoted to the Spanish writer Cervantes is located in the writer's house in Alcalà de Henares, near Madrid.

In designing such a museum (or exhibition), it is important to be as objective and *aseptic* as possible, by avoiding any hagiographic aspects. It is very opportune that the exhibition be complemented by exposition of the individual's personal objects, such as original books, papers, and everyday objects.

Appropriate background noises associated with the person could also be very significant. As an example, an exhibition held in Pavia in 2007 and devoted to the great zoologist Pietro Pavesi had, as acoustic background, the *voices* of animals.

In this category can be included museums located in the homes of noble families, with furnishings, furniture, objects that belonged to the family, etc. These are interesting collections documenting the life of a family and an era. Two examples are the Museum Poldi Pezzoli and the Museum Bagatti Valsecchi in Milan.

5.7.12 Technological Processes

A technological process (i.e., the process that starts with a raw material and ends up with a finished product) can be presented in the following steps:

- starting material
- working activities
- machines and plants
- finished product.

Such steps should be presented highlighting the historical evolution of the most important examples and, of course, through reproduction of the psychological state of the workers who participated in the process described.

5.7.13 Foods

Foods can be presented either through the ways in which they are processed (machinery and production plants) or in regard to their standard use. For example, the following ideas can be useful:

- Overview of historical and modern agricultural machines, with their constructive evolutions highlighted.
- Constructive details about agricultural machines, with explanations of the agricultural processes to which they relate.
- Cycles of processing of food packaging.
- Videos featuring people giving explanations and/or demonstrations on the preparation of foods.
- Historical evolution of household appliances (with constructive details highlighted) and the types of food for which they are useful.
- Shelves of a store with the products on display.
- A table spread.

In some cases (for example, specific exhibitions), tastings of typical products may be offered.

5.7.14 Reproduction of Scents

In some cases, the reproduction of scents can be useful, e.g., in a museum devoted to the kitchen or a particular kind of food, or simply to complement the reconstruction of a situation or an environment devoted to a specific activity. The scents can be reproduced through the use of containers filled with extracts of odorous substances.

5.8 Explanations

During a visit to a museum, explanations are very important, because they represent a significant upgrade in the degree of knowledge that the visitor can obtain. Explanations can be presented in many forms:

(a) by a real person. The person who delivers these explanations can be an authorized external expert or a member of the scientific staff of the museum. In some cases, the person who explains can be an actor playing the role of a

scientist, perhaps wearing clothes corresponding to the time, presenting scientific principles and laws, or technical realizations. Of course, the level of the explanation must be congruent with the level of the audience, taking into account their "initial behavior" and the "terminal behavior," as explained in Chap. 2. From this point of view, it is interesting to note that accommodations must be made so that the explanations are also accessible to disabled persons. As an example, people with poor eyesight can explore models of objects through tactile perception, of course, with adequate explanation
(b) by a virtual person. The "virtual professor" in a (real!) environment can, from within the display itself, give explanations and quiz the visitors. The visitors can answer the questions through use of a device on a desk and the virtual professor can then verify whether their answer is right or wrong
(c) with static video
(d) with registration related to the succession of the expositive sections
(e) with 3D Viewers: more than simply providing explanations, these can truly complement the visit, making possible a virtual experience in which there are no physical objects; the visitor is able to have otherwise impossible experiences in the imagined reality, such as passing through walls and moving through an unreal space.

5.9 Captions

Some criteria for writing captions are as follows:

(a) A caption must always be adjacent to the object it describes;
(b) The size of the fonts should be readable at a distance not very different from that required to observe the object;
(c) The position must be chosen so as to facilitate reading by both adults and children: around 1 m in height is generally suitable;
(d) The text must be in two languages: the language of the country where the museum is located and English;
(e) The text must be very complete and exhaustive (obviously!), including all necessary information, or at least, all useful information. It could be interesting to lend the captions a hypertextual character, e.g., general information for all visitors, and more specific detailed information for those most interested.

The above-mentioned categories of information can be highlighted, either in different fields or with different colors for the captions. Figure 5.2 depicts the general schema of captions, with three levels of depth. The items in a general caption should be as follows:

(a) name of the object
(b) author

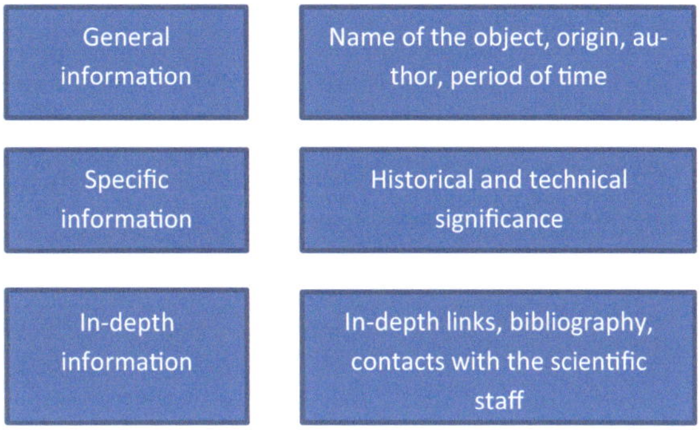

Fig. 5.2 Example of categories of information to lend the captions a hypertextual character

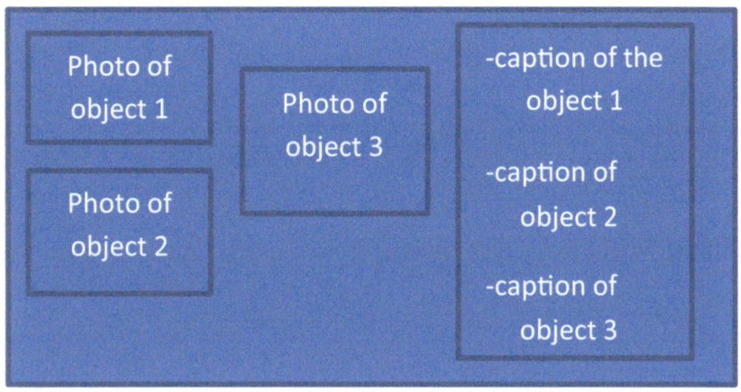

Fig. 5.3 Captions related to many objects, using photos of the objects as reference

(c) year/period of time
(d) source
(e) description (this particular aspect can, of course, be given a further level of depth if you want to differentiate certain specific types of information according to the different interest levels of the visitors).

For placement of captions, the most important possibilities are as follows:

(a) adjacent to each exhibited object: in this case, it is generally easier to read the text and there is the possibility of making the caption shorter and more succinct;
(b) only one caption related to the placard in each room or each showcase: in this case, the text is easier to realize but more difficult to read, particularly because it requires more time to connect the text with the corresponding object. All the captions are united in a table or a placard, which should represent one of the following configurations;
(c) only captions;
(d) captions and photos of the exhibited objects;
(e) captions and identification numbers of the corresponding objects (Fig. 5.3).

From a communication point of view, solution (d) seems to be the most useful.

- If the object is illuminated, the captions should also be illuminated.
- Digital captions can also be very interesting, as their dynamic presentation can lend greater depth to the information or present the explanation in more languages.
- As for captions, they could be presented on an informative sheet available for consultation by visitors, located in a special container in each room.

More information is available in Cataldo and Paraventi (2007).

References

Biggioggero GF, Calabrò S, Menzio G, Rovida E (2003) Evoluzione storica di soluzioni costruttive. Il caso di sospensioni per autovettura Politecnico di Milano—Dipartimento di Meccanica CNR (National Council of Researches) Progetto finalizzato Beni culturali

Cataldo L, Paraventi M (2007) Il museo oggi. Linee guida per una museologia contemporanea Hoepli, Milano

Chapter 6
Some Auxiliary Communications

6.1 Introduction

The museum (and, respectively, the exhibition) is a very complex form of communication, constituted by an integration of many more elementary forms of communication.

Museums and exhibitions make wide use of auxiliary communication, such as tickets, free documentation and catalogs; such auxiliary communication can play an important role in the formation of cultural cooperation with museums and exhibitions.

In fact, they can give information complementary to the exhibited placards, objects and captions. In addition, such auxiliary means of communication can be retained by the visitor, becoming a sort of physical "memory".

6.2 Tickets (Mazzanti 1996)

The ticket has an administrative function; it can, however, have an additional cultural function, of interest to the visitors. In this way, the ticket can be an interesting means of communication. In some cases, the ticket can also have historical and collectible interest.

There is a very wide variety of tickets, related to the formal aspect of the exhibition and the general tone of the information presented.

Among such a wide variety of tickets, some categories are recognizable as follows:

(a) basic tickets. Such tickets have the name of the museum, sometimes, but not always, the price of admission, and a progressive number
(b) tickets with the logo of the museum. Such tickets have the above-mentioned information and, in addition, the logo of the museum, often as a result of a deep

A. Rovetta and E. Rovida, *Scientific Knowledge Communication in Museums*,
Research for Development, https://doi.org/10.1007/978-3-319-68330-0_6

graphics research, and possibly some complementary information related to the museum, e.g., closing days, list of the individual sections with museum hours, etc.

(c) tickets as magnetic cards. Such tickets, often enhanced with aesthetically pleasing pictures related to the exhibited collections, have, on the other side, a magnetic strip, so as to control entrances
(d) tickets configured as labels. These tickets are configured as labels, with a picture related to the exhibited objects and/or some essential information. On the other side, they have an adhesive that allows them to be attached to the clothes of the visitor
(e) tickets configured as pins. This ticket is in the form of a pin to be placed on the clothes of the visitor. The pin has a picture, with, e.g., the logo of the museum
(f) tickets with deeper information. Such tickets are characterized by the presence of more in-depth information about the museum, e.g., a map, a list of sections, a list of the most important exhibited objects, museum hours, schedule of days the museum is open, information about possible paths, etc.
(g) tickets configured as flyers. Such tickets are constituted by the flyer of the museum, marked by a stamp with, e.g., the day of the visit, the number of the visitor, the price, etc. The flyers can contain a lot of important information about the museum, and such tickets are, consequently, similar to small catalogs.

Figure 6.1 depicts a ticket for the Lombard Museum of the History of Agriculture; the ticket is of a traditional form, with a very simple logo, the name and location of the Museum, and the progressive number.

Figure 6.2 represents four examples of tickets from the Museum "Leonardo3—Il Mondo di Leonardo" of Milan "The Da Vinci's Museum of Milan." Such tickets are modern and the result of a careful graphic study.

After observation of many different types of ticket, it is possible to synthesize a list of the most important pieces of information that can appear on them:

- name of the museum;
- logo of the museum;
- number;
- a picture that characterizes the museum (e.g., a significant object);
- general information about the museum (e.g., address, how to reach the museum, opening hours);

Fig. 6.1 Example of a traditional ticket (courtesy of MULSA (Lombard Museum of the History of Agriculture) of S. Angelo Lodigiano, Lombardy)

Fig. 6.2 Four examples of tickets featuring pictures created by Leonardo da Vinci (courtesy of "Leonardo3—Il Mondo di Leonardo" of Milan)

- specific information about the museum (e.g., description of the museum, sections, its most important objects).

In some cases, as mentioned earlier, the ticket can have enough information contained within it so as to qualify as a small catalog.

The physical configuration of the ticket can also be quite varied:

- paper (card, postcard, pamphlet);
- magnetic card;
- label;
- pin;
- stamp (on the hand of the visitor!).

6.3 Free Documentation

By "free documentation," we mean sheets, flyers and other printed matter with general information about the museum or about a specific section or room. Such documentation is free for the visitors. In the great variety of such documents, some general types are recognizable.

(a) a sheet with a characteristic picture: this is a piece of paper with the museum logo and essential information;
(b) a sheet with a map of the museum/exhibition: can include suggested paths, often highlighting the most interesting objects on exhibit;
(c) a flyer with general information about the museum/exhibition (days and hours, short descriptions of prominent exhibitions);
(d) a flyer with a detailed map, list of sections and rooms, and basic information;
(e) a flyer with a detailed map, list of sections and rooms and more detailed information;
(f) a sheet with a list of events, e.g., exhibitions, conferences, courses organized by the museum.

Utilization of free documentation can be, in general, as follows:

- purely for consultation during the visit: in many museums, each room or section has a container full of sheets to consult, with information about the room/section and the most important objects exhibited;
- to take home as a souvenir: this is very similar to the case (a), with the difference being that the visitor may take the sheet home and keep it as remembrance of the visit. Most of the above-mentioned sheets from (a) to (f) belong in this category.

6.4 Catalog

The catalog can be defined as a document, paper or digital, with the aim of transmitting information to the visitors, facilitating the visit and also reinforcing and deepening what they have learned during the visit. Catalogs can have very different configurations, in relation to the information presented and the physical configuration.

Table 6.1 presents a general list of the sort of information that could be contained in a catalog. Of course, in relation to the objectives and configuration (and cost!) of the catalog, the items can be reduced as needed.

The above-mentioned parts can be combined and organized to create very different configurations of catalogs, from a simple leaflet to a scientific treatise. Below, some pages of catalog are presented, corresponding to various configurations.

Table 6.1 General list of information in the catalog

General sections	Specific sections
1. Introduction	1.1. General introduction 1.2. Presentation of the sponsors 1.3. Presentation of the scientific committee
2. Location	2.1. Address 2.2. How to reach the museum 2.3. History of the museum 2.4. Description and history of the building where the museum is located 2.5. Choice of the location 2.6. Maps of the museum 2.7. General criteria of the visit
3. General part	3.1. General monograph/s related to the mission of the museum and the general characteristics, such as, e.g., the nature of the exhibited objects
4. Specific parts	For each section (a chapter of the catalog): 4.1. Specific monograph related to the section 4.2. Specific catalog of the section (list with photos and descriptions of the exhibited objects, or at least of the most significant, and reproduction of the exhibited placards) 4.3. Specific bibliography
5. Conclusion	5.1. General bibliography 5.2. Chronological tables 5.3. Citations and press reviews 5.4. List of related museums 5.5. List of the exhibitions organized by the museum 5.6. Associations of the collectors of the objects exhibited in the museum 5.7. Biographical notes 5.8. Glossaries 5.9. Scientific institution/s located in the museum 5.10. Alphabetic index 5.11. Analytical index

The general contents presented in the table, through different combinations and extensions, are the basis for certain configurations of catalog, e.g.:

(a) basic, with only the most important information and with a reduced extension;
(b) essential, with a greater degree of information;
(c) complete, with copious photos and descriptions of the most important objects exhibited;
(d) monograph, with significant documentation about the exhibited objects and an extensive monograph about the historical and cultural role of the exhibited collections.

From the point of view of physical configuration, it is possible to distinguish certain forms of catalog:

- paper, in the form of leaflets, dossiers, booklets, books, treatises;
- audiovisual (now rarely used);
- digital (e.g., in the form of a USB key);
- available on the website of the museum.

Figures 6.3 and 6.4 depict the catalog from the "History of Drawing" Museum (Brunetti and Rovida 2006). This type of catalog is constituted by photos of the exhibited instruments with a short description of each of them.

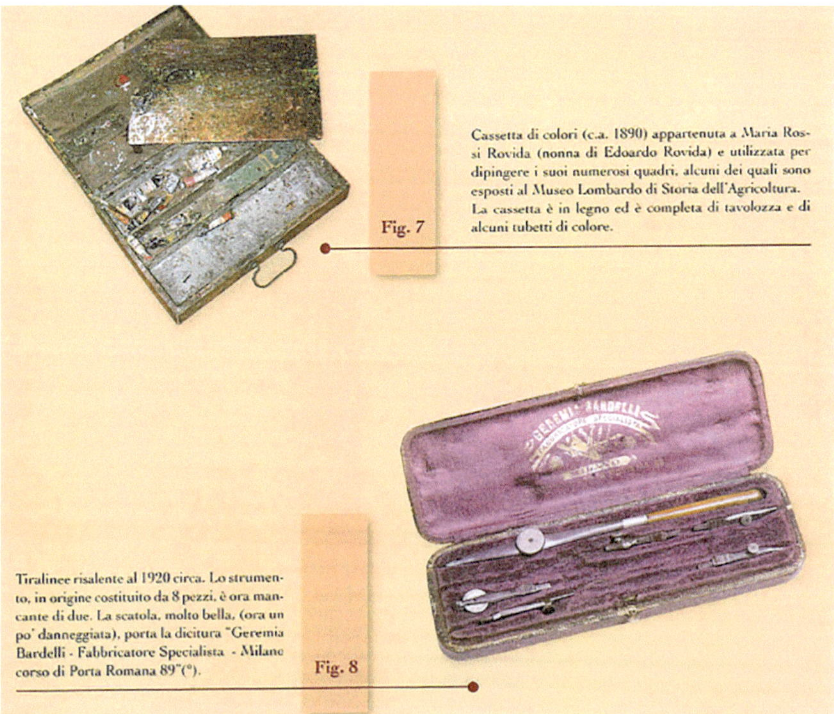

Fig. 6.3 Drawing instruments with short descriptions

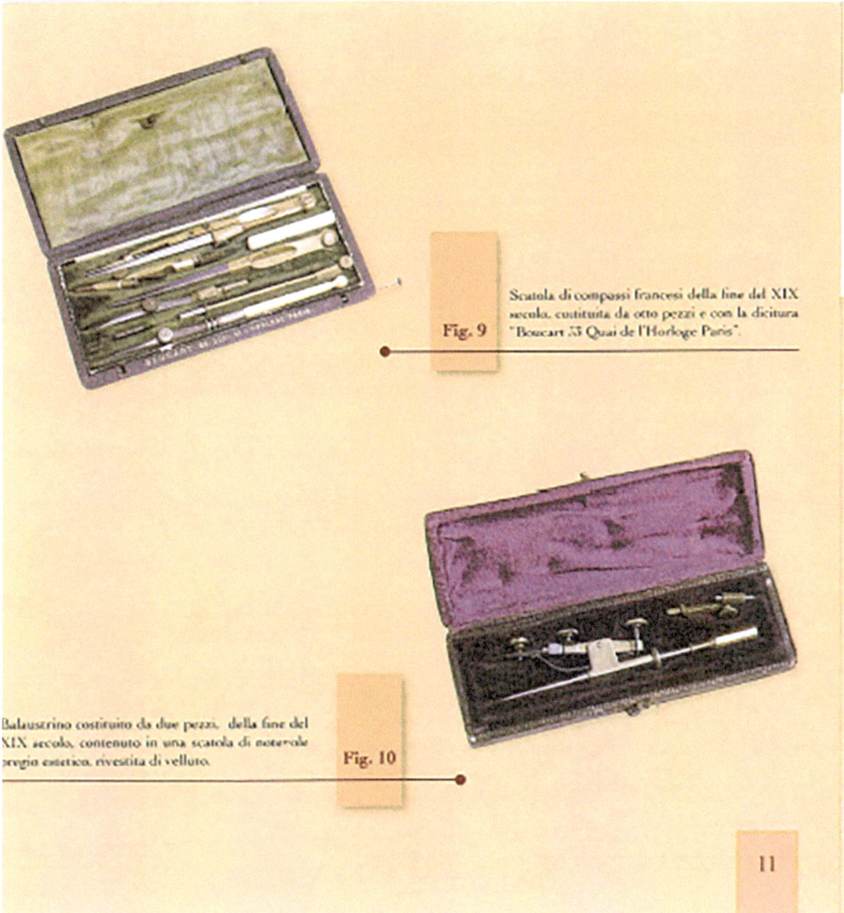

Fig. 6.4 Drawing instruments with short descriptions

Figures 6.5, 6.6 and 6.7 depict pages from the catalog of the Lombard Museum of the History of Agriculture (Mariani 2014) in S. Angelo Lodigiano.

The figures represent a particular integration of photo and description: the description is sufficiently broad, and therefore can be considered a kind of monograph.

The page shown in Fig. 6.5 is related to plows, while Fig. 6.6 depicts moments from the agricultural life of the ancient Romans, expressed through bas-relief. The page in Fig. 6.7, for its part, is all about agricultural machines (threshers).

Another type of catalog configuration is constituted by reproductions of all the placards presented throughout the exhibition. Figure 6.8 depicts such an example (aa.vv 1988).

Strumenti agricoli e vasellami originali (reali o simbolicc rituali) tratti dai vari continenti

OGGETTI IN PRESTITO DAL MUSEO NAZIONALE PREISTORICO ETNOGRAFICO "LUIGI PIGORINI", ROMA

Il Museo Nazionale Preistorico Etnografico è un'istituzione pubblica al servizio della società multiculturale e del suo sviluppo civile che fa ricerca, assicura la salvaguardia e promuove la comunicazione, a fini conoscitivi, educativi e ricreativi, del patrimonio materiale e immateriale d'interesse palotnologico ed etnoantropologico.

Il Museo Pigorini, fondato nel 1875, ha sede a Roma E.U.R. in piazza Guglielmo Marconi, 14, è ordinato in due grandi settori espositivi permanenti dedicati all'Etnografia e alla Preistoria.

Quattordici oggetti della sezione Etnografia sono esposti presso il nostro Museo per testimoniare la matrice tecnologica e culturale comune dell'agricoltura dei nostri diversi continenti.

ASIA

1 - Modellino di aratro rituale
Funzione: aratura
Origine: Thailandia

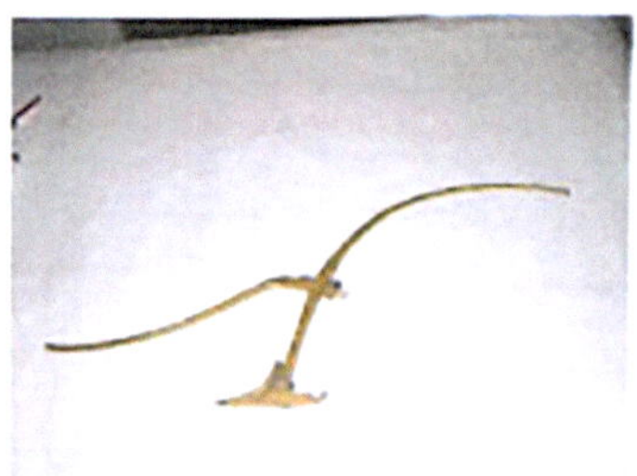

2 - Modellino di aratro rituale
Funzione: aratura
Origine: Thailandia

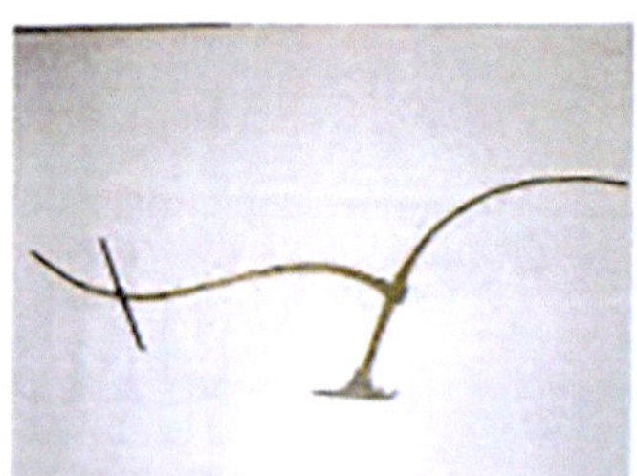

15

Fig. 6.5 Example of a page from the catalog of the Museum of the History of Agriculture (courtesy of MULSA (Lombard Museum of the History of Agriculture) of S. Angelo Lodigiano, Lombardy)

Riproduzione di bassorilievo raffigurante il rito di fondazione di una città: si delimitano le mura esterne mediante un solco tracciato con un araro trainato da buoi (Museo Archeologico di Aquileia, I secolo a.C.).

Da queste e da altre opere, di cui abbiamo solo frammenti, ricaviamo notizie non solo sulle piante coltivate, gli animali allevati, gli strumenti usati, ma anche sulle operazioni colturali, l'organizzazione dell'azienda agricola, la specializzazione in essa della mano d'opera, la conduzione economica, con minuziose annotazioni di costi e rendimenti.

A sinistra riproduzione di bassorilievo raffigurante la mietitrice gallica dei primi secoli d.C. (Montauban-Buzenol Francia, I secolo d.C.); sotto lo stesso soggetto nella ricostruzione grafica di H. Cüppers, 1964.

Fig. 6.6 Example of a page from the catalog of the Museum of the History of Agriculture (courtesy of MULSA (Lombard Museum of the History of Agriculture) of S. Angelo Lodigiano, Lombardy)

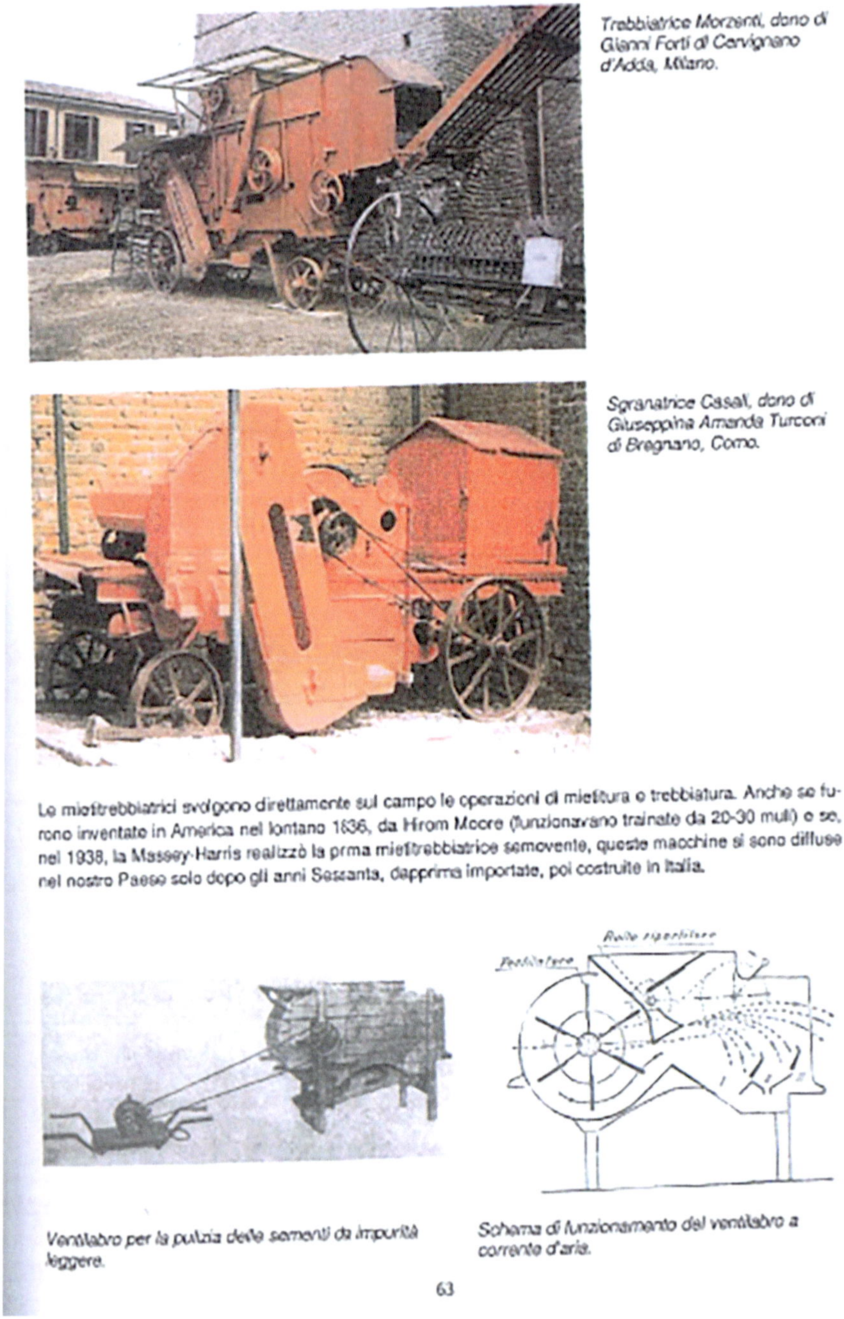

Trebbiatrice Morzenti, dono di Gianni Forti di Cervignano d'Adda, Milano.

Sgranatrice Casali, dono di Giuseppina Amanda Turconi di Bregnano, Como.

Le mietitrebbiatrici svolgono direttamente sul campo le operazioni di mietitura e trebbiatura. Anche se furono inventate in America nel lontano 1836, da Hirom Moore (funzionavano trainate da 20-30 muli) e se, nel 1938, la Massey-Harris realizzò la prima mietitrebbiatrice semovente, queste macchine si sono diffuse nel nostro Paese solo dopo gli anni Sessanta, dapprima importate, poi costruite in Italia.

Ventilabro per la pulizia delle sementi da impurità leggere.

Schema di funzionamento del ventilabro a corrente d'aria.

63

Fig. 6.7 Example of a page from the catalog of the Museum of the History of Agriculture (courtesy of MULSA (Lombard Museum of the History of Agriculture) of S. Angelo Lodigiano, Lombardy)

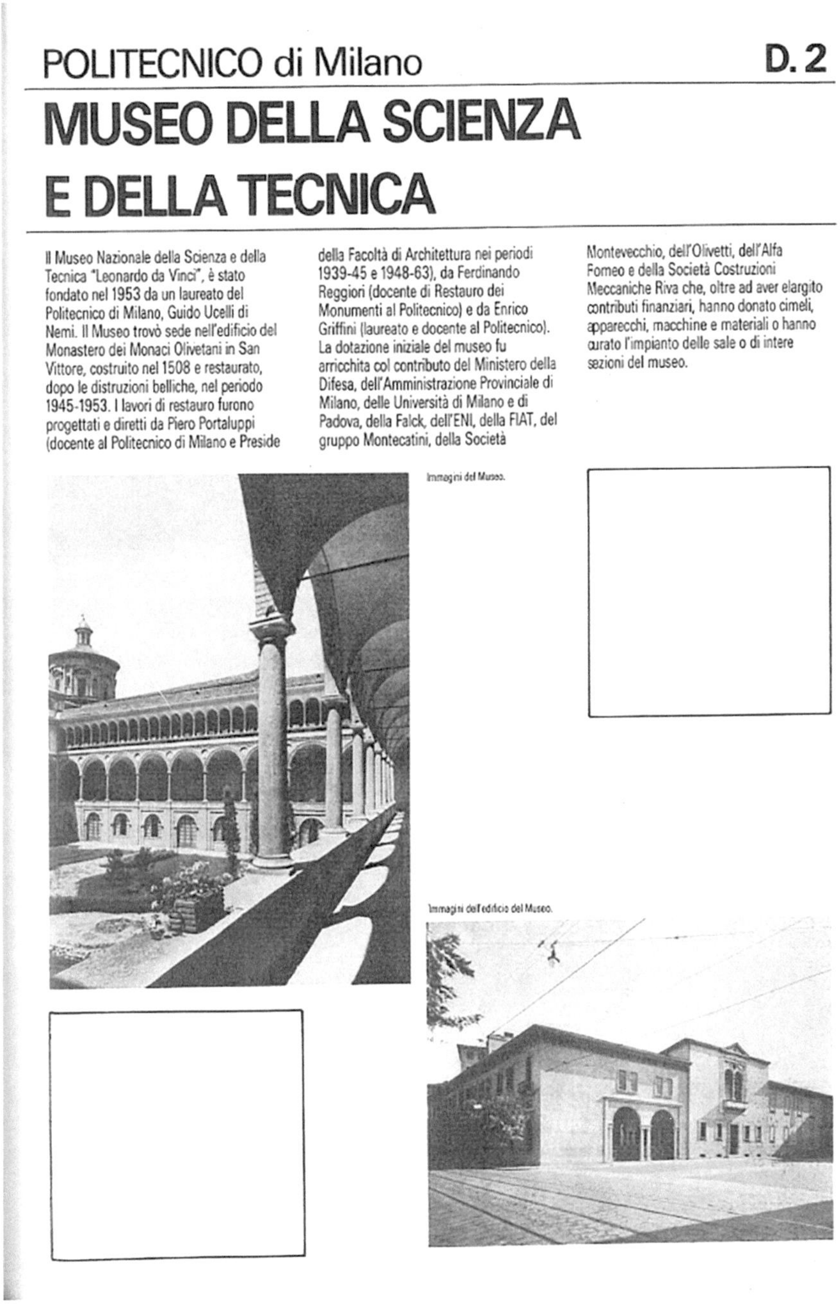

POLITECNICO di Milano D. 2

MUSEO DELLA SCIENZA E DELLA TECNICA

Il Museo Nazionale della Scienza e della Tecnica "Leonardo da Vinci", è stato fondato nel 1953 da un laureato del Politecnico di Milano, Guido Ucelli di Nemi. Il Museo trovò sede nell'edificio del Monastero dei Monaci Olivetani in San Vittore, costruito nel 1508 e restaurato, dopo le distruzioni belliche, nel periodo 1945-1953. I lavori di restauro furono progettati e diretti da Piero Portaluppi (docente al Politecnico di Milano e Preside della Facoltà di Architettura nei periodi 1939-45 e 1948-63), da Ferdinando Reggiori (docente di Restauro dei Monumenti al Politecnico) e da Enrico Griffini (laureato e docente al Politecnico). La dotazione iniziale del museo fu arricchita col contributo del Ministero della Difesa, dell'Amministrazione Provinciale di Milano, delle Università di Milano e di Padova, della Falck, dell'ENI, della FIAT, del gruppo Montecatini, della Società Montevecchio, dell'Olivetti, dell'Alfa Romeo e della Società Costruzioni Meccaniche Riva che, oltre ad aver elargito contributi finanziari, hanno donato cimeli, apparecchi, macchine e materiali o hanno curato l'impianto delle sale o di intere sezioni del museo.

Immagini del Museo.

Immagini dell'edificio del Museo.

Fig. 6.8 Example of a page from the catalog of the exhibition celebrating the 125th anniversary of the founding of the Politecnico di Milano

The figure presents a page from the catalog of the exhibition celebrating the 125th anniversary of the founding of the Politecnico di Milano.

Of course, the placards themselves were realized taking into account the two specific functions they would be called on to perform: dispensing information during the exhibition and representing that information as reproductions in the catalog.

References

aa.vv (1988) Il 125° del Politecnico di Milano Cittastudi Milano

Brunetti R, Rovida E (2006) La Sezione Disegno Tecnico del Museo Lombardo di Storia dell'Agricoltura, Lettera 22, Pantigliate

Mariani L (2014) Guida alla visita ragionata Museo Lombardo di Storia dell'Agricoltura (Lombard Museum of the History of Agriculture). Available on https://sites.google.com/site/storiagricoltura/museo. Last visit 12 Mar 2017

Mazzanti (1996) Il museo dei biglietti dei musei. Amici dei Musei 6–9:42–43

Chapter 7
Role of the Museums in the Scientific-Technical Education

7.1 Introduction

It is important, if regrettable, for us to acknowledge the fact that science and technology play a very small role in the general culture today. Scientific-technical culture is often considered only on the basis of its ability to solve practical problems, as a tool. However, science and technology, as mentioned in Chap. 1, play a fundamental role in the life of each person nowadays (including for people active in humanistic professions). The lack of an adequate scientific-technical culture can very often cause serious, even catastrophic, problems.

This idea is not put forward for the purpose of minimizing the importance of humanistic culture, which is itself very significant. Scientific-technical and humanistic cultures are both important, indeed complementary: humanistic culture alone, without a scientific-technical component, is essentially just erudition of stories from our past, while scientific-technical culture can be considered to be nothing more than training in a particular discipline without a humanistic component. Museums, and particularly scientific-technical museums, can (and must) provide important contributions to scientific-technical education.

7.2 Links Between Museums and Scientific-Technical Education

The links between museums and scientific-technical education are copious and significant. A general schema of the above-mentioned links is presented in Fig. 7.1.

Scientific-technical education can be distinguished in two forms: institutionalized (i.e., courses taught at Universities and/or official educational Institutions) and

A. Rovetta and E. Rovida, *Scientific Knowledge Communication in Museums*, Research for Development, https://doi.org/10.1007/978-3-319-68330-0_7

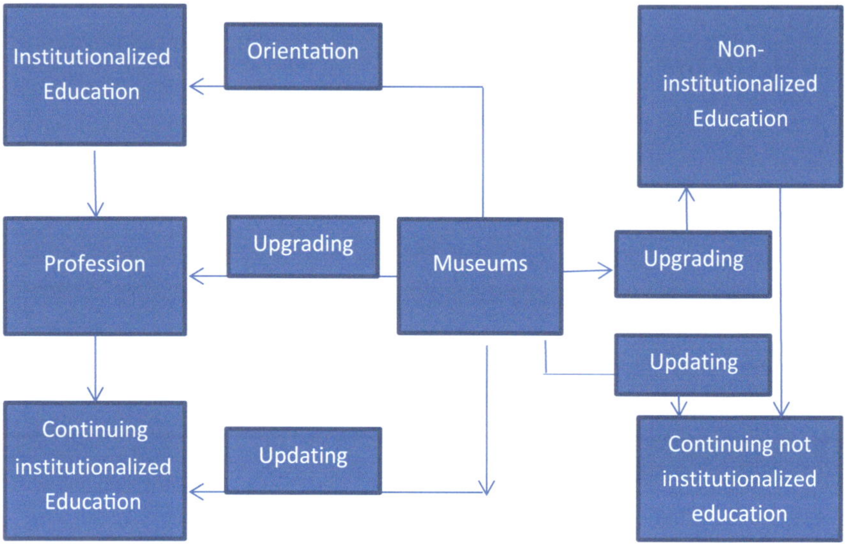

Fig. 7.1 General schema of the links between museums and scientific-technical culture

non-institutionalized (i.e., realized through reading books and papers, participation in conferences, autodidactic studies).

Three fundamental activities strictly related to education can be recognized as follows:

(a) orientation, which relates to information dispensed before decisions affecting future activities, e.g., the type of school to attend or a desired professional activity;
(b) upgrading, which refers to a deeper and/or more comprehensive form of information dispensed during educational activities;
(c) updating, which refers to the dispensation of more recent information during, or after, educational activities, with the aim of avoiding the obsolescence of the acquired knowledge.

7.3 Contribution of Museums to Scientific-Technical Education

The contribution of museums to scientific-technical education can be examined in relation to the above-mentioned points of view: orientation, upgrading, updating.

7.3.1 Orientation

Orientation can take place, in general, in the following moments:

(a) before (during) primary school
(b) before (during) secondary school
(c) before choice of a university
(d) before choice of a profession.

Such situations are now considered:

(a) before (during) primary school. In this case, the educational function of museums is directed at children: therefore, the language and means of communication should be congruent with the above-mentioned target. The presentation of the exhibited objects (and, particularly, of the scientific-technical objects) must take into account the important role of the emotional aspects of the objects themselves. For example, it is useful to stimulate the curiosity of the children by linking the exhibited objects to their own lives (home, family, play, toys, and so on). This objective can sometimes be reached through differentiation of the itineraries. Such differentiation could simply be related to portions of the captions, examples, pictures, etc. With good presentation, the interest of children in science and technology can be upgraded. The experience of the Author is as follows: during a visit in 1953 to the Museo Nazionale della Scienza e della Tecnica (now Tecnologia) (the National Science and Technology Museum), the Author, at the time a child and student in primary school, became fascinated by an exhibition celebrating the 500th anniversary of the birth of Leonardo da Vinci. On that occasion, most likely, the Author's vocation in mechanical engineering was born.
In addition, laboratories can be organized where children, through play, can learn about principles of science; for example, in regard to physics, by playing on a seesaw, the children can discover the principle of the lever
(b) before/during secondary school. The children are now grown and have greater and deeper cultural knowledge and interests. Their itineraries can be deeper as well, and the possibility of conducting experiments is increased. The laboratories can become more specialized, and there can be more of them, allowing for the possibility that they might conduct their own experiments autonomously. Such experiments could be conducted in either of the following two fundamental ways:

- a physical system contained in a case: by pressing a button, the experiment starts
- the system is presented through a virtual model and the visitor acts, virtually, on said model.

In addition, conferences could be organized, with the aim of presenting more comprehensive information to the students.

(c) before university. The choice of institution could be facilitated through deeper itineraries, such as information on specific technologies or work environments (the emotional aspect of mining, workshops, factories), which can provide valuable input to the students before their choice. In such moments, conferences, guided tours through specific museums, and contact with the scientific staff of the museum can be useful.

7.3.2 Upgrading

The contribution of museums to the upgrading of scientific-technological education can be distinguished in relation to the intended audience: students or teachers.

7.3.2.1 Upgrading for Students

Students in primary and secondary schools, as well as universities, can, through museums, upgrade their education from many points of view.

The students can gain advantage through informed exposure to the objects and exhibits presented in the museum. From this point of view, the students can see and study instruments, machines and devices that they would never be able to see in school. Such objects, machines, instruments and devices can be presented in their historical perspectives, according to the well-known advantages of history, highlighted in Chap. 1. In addition, the students can “walk” through a working environment and, consequently, can experience the mood and sensations of the people who work (or used to work) in such environments. Of course, the presentation is an important responsibility of the curator, who must be attentive to certain essential aspects:

- acquisition and, of course, exposition of the greatest possible number of significant machines, instruments and devices, representative of a wide range of constructive solutions;
- presentation of such machines, instruments and devices with all the necessary explanations, such as drawings, pictures, functional schema, texts, advertising and all other documents that might complete the description of the objects;
- such objects, as mentioned above, should be presented firmly within their historical perspective;
- if real objects are not available, pictures, drawings and virtual models could be useful;
- such objects should be presented by highlighting the emotional aspects that, as has been observed by the Authors on many occasions, can play a role that is very important to and quite interrelated with, from an educational point of view, the logical and rational aspects. Such aspects that the exhibited objects might

conjure up as love, consciousness, pain, emotion, time, participation, and pleasure should be highlighted.

- the configuration of the museal itineraries could also have great importance: for example, a differentiation of specific itineraries related to the level of school can be studied and organized.
- laboratories are, likewise, very important for the purpose of testing, confirming, and repeating the knowledge acquired in school;
- another interesting possibility is the involvement of students as guides for the visitors, giving explanations during the visit. The Author has some students who are involved in an important museum in Milan. In the course "Communication Skills," held at the Politecnico di Milano by the author, these students presented, as an exercise, some critical observations regarding the above-mentioned activity.

7.3.2.2 Upgrading for Teachers

Teachers can also be usefully involved in museal activity, upgrading their own professional preparation in the process. Some examples are as follows.

(a) Consulting during the design and realization of a new section (or the redesign/restoration of existing sections) of a museum, in relation to their specific field of expertise. Similarly, the scientific competences of teachers could be useful for the design and realization of specific exhibitions.

(b) Disposition of teachers as contacts for specialized visitors. Visitors with a much deeper degree of scientific interest, such as actual professional scientists, wishing to expand their knowledge could contact the scientific staff of the museum. In this respect, cooperation between curators and teachers could be very useful.

7.3.2.3 Upgrading for Teachers and Students

The upgrading of the cultural and social function of the museum can result in an upgrading of existing professions and the creation of new ones.

As an example, some possible new or upgraded professions could appear in the following fields:

- disciplines related to cultural heritage
- informatics, necessary for museum management
- economics and management sciences
- tourist sciences
- museum curation

- languages
- use of 3D printers and holography, to create gadgets to be distributed in bookshops.

7.3.3 Updating

During updating, the contribution of scientific-technical museums can also play a significant role. As mentioned above, updating involves maintaining the validity and modernity of acquired knowledge. The activity of updating is often called "Continuing Education."

The fundamental purpose of continuing education is to maintain adequate skills acquired over time, known as 'terminal behavior' (see Chap. 2). The required knowledge, starting from the doctoral degree and continuing on through retirement, must be continuously upgraded, in relation to increasing professional requirements. Continuing Education aims to upgrade the knowledge of the individual.

Institutionalized Continuing Education tends to be concentrated in correspondence courses, workshops, and seminars, organized by official Institutions. Non-institutionalized Continuing Education, however, focused as it is on reading books and scientific papers in a purely individual capacity, is less concentrated and more distributed over time.

Museums can play a role in both of these forms of education.

In institutionalized education, museums can cooperate with other official Institutions, by organizing conferences, seminars, and workshops, in this case, associated with a system of credits.

In non-institutionalized education, the role of museums can be similar (but, of course, without credits) to that mentioned above.

In addition, by upgrading the exhibited objects and the criteria of exposition, a visit to the museum can be a significant occasion for the upgrading of knowledge.

With reference to the upgrading of the exhibited objects, a significant problem, particularly, but not exclusively, in the case of scientific-technical museums, is the upgrading of the collections, which always takes more time than scientific-technical evolution. An interesting solution was discovered, many years ago, by the Museo Nazionale delle Scienza e della Tecnica (now "Tecnologia") "Leonardo da Vinci" (the National Museum of Science and Technique (now "Technology") "Leonardo da Vinci") in Milan. The solution, called a "Salone dell'Innovazione" ("Innovation Exhibition"), was a presentation of the most recent and innovative technical products, accompanied by explanations, placards and captions. The exhibition was held two times during the year: a visit was an important occasion to see and learn about the most innovative products in many different fields, and therefore an occasion for upgrading professional preparation.

Today, the most modern solutions can be realized through informational applications.

7.4 Conclusions

Museums, if properly designed, realized and managed, can play an important role in cooperation with educational Institutions, particularly in the scientific-technical field.

The present chapter presents certain examples of such cooperation: of course, such examples are not exhaustive, but can be upgraded and enlarged, through the creativity of teachers and museologists.

Some thoughts on the teaching aspects of museums, with particular attention to scientific-technical museums, can be found in Rossetti and Sassi Perino (1989), Rovida (1994, 1995) and Cataldo and Paraventi (2007).

References

Cataldo L, Paraventi M (2007) Il museo oggi. Linee guida per una museologia contemporanea Hoepli Milano

Rossetti U, Sassi Perino AM (1989) The role of the historical museum of the Politecnico di Torino in education and in the conservation of the local cultural heritage Paper presented at the SEFI (European Society of Engineering Education) Annual Conference, Naples 17–20 Sept 1989

Rovida E (1995) The role of science museums in design education Paper presented at the International Conference on Engineering Design ICED 95 University of Praha, 22–24 Aug 1995

Rovida E (2004) The role of science and technology museums in engineering education Paper presented at the symposium IGIP (Internationale Gesellschaft fuer Ingegneurpaedagogik, International Society for Engineering Education) University of Praha, 21–23 Sept 1994

Part II
Dynamic Museums and Social Learning

Chapter 8
Social Learning, Around 2020

The Creation of Learning Is as Old as the Birth of Human Intelligence
In the thousands of years during which what we call civilization has existed, knowledge and learning have traveled hard roads and achieved very profound results. In 360 BC, Aristotle had already indicated the lines, traced till today, by the AI, Artificial Intelligence, because he wrote first about "qualities" as appearances of reality. 3500 years ago, the Egyptians were building high-tech items. Knowledge through learning was exclusive and intimate. Only 50 years ago, schools were very selective and preserved knowledge for those lucky enough to have access to it. Today, approaching 2020, information moves like lightning, and both knowledge and 'know how' are available to billions of people.

A. Rovetta and E. Rovida, *Scientific Knowledge Communication in Museums*,
Research for Development, https://doi.org/10.1007/978-3-319-68330-0_8

Courtesy of Collection AR

Life emotions colors

A SHIP IN MOTION, A BRIDGE FOR STANDING STILL, AN OCEAN FOR DEPTH, FROM THE WINDOW OF A MUSEUM

In 1957, to collect the sum total of engineering knowledge would take at least a day. Today, 50 years later, it takes, assuming the computer is on, no more than 60 s, and with a far greater range of information completely available. Today, the mind seeking knowledge through social learning has a whole world in which to participate, in ways that are both easier and, in general, positive.

Our reality is the world made up of our experience, and the source of experience is the brain, which has evolved, while physical reality has been built up over time by different civilizations, in which each of us lives. We are all connected within a social world, which is understood as socius, from the ancient Latin, meaning "participant, cooperative, supportive."

Human cultures have led to human interpretations of the world, and the human brain can recognize the fact that we are now capable of expressing ourselves through both virtual reality and augmented reality.

Art, science, culture, reason and emotion are all facets connected to the human process as a whole.

Nature, history, the world and the mind are co-stars in the social learning of modern life. To a more and more intense degree, reality is becoming a human creation.

In this project, we are building a system for social learning through the reality of the museum. It is necessary to consider that in museums, there is an unspoken poetry, and that we must turn it into music, words, and images. Each object has its own sensitivity and intelligence, and it must become a talking piece, using those words, those poems, those notes to bring itself closer to your experience, your culture and your existence of mind and body.

8.1 Hymn to the Cultural Heritage in Our Goals

Welcome to the presentation of the Dynamic Museums (DM) project, a project we hope will be noted for its positive approach to the museums of Italy and the world.

The DYNAMIC MUSEUMS Project currently runs on tablets and electronic devices, on a digital platform, meaning that, through software and hardware connected with the databases of various museums, all Italian for now, the visitor gets to build a museum path based on his/her immediate state of mind. The visitor clicks on the values related to that state of mind, using both logic (left side of the brain) and emotion (right side of the brain), and the platform, accessing all objects within the museum, as well as the objective evaluations of the curators of the museum, builds a museum visit path, adapted to the visitor in his/her present state of mind, doing so in mere millionths of a second.

There are examples of such museum paths, one dedicated to Milan, and another one dedicated to an Italian Archaeological museum, included in the package of agreement between the Ministry of Cultural Heritage and Activities and Tourism of Italy, the Fondazione Politecnico di Milano with the Department of Mechanics of the Polytechnic, and the DYNAMIC MUSEUMS Project.

DM stands as a small contribution towards the idea of a better world, one that has gained emphatic support, particularly after the trial run in Italy and the invitation from the ICOM World Federation of Museums to collaborate with them.

DM is a hymn to the joy of the present, with the hope of improving peoples' love towards the present and towards themselves.

It is a hymn to love in real life, in its variegated colors, light and dark.

It is a hymn to mutual experience, along various routes, with infinitely different paths.

It is a hymn to the strong and beautiful memories of lived hours, related to the objects within the museums of the world.

It is a hymn to the soft power of thought that curls the objects into the circular map of memory.

It is a hymn to the dream and the desire of the future, full of the goodness leavened in the DM project.

We explain and show the "game", which is the first in the world that uses emotions that we know it can procure.

More than 1600 witnesses in Italian museums, beginning on August 2, 2016, have used and are using it.

Donated by the Ministry of Cultural Heritage and the Fondazione Politecnico di Milano, both of whom have given us their great support, along with the patronage of the Mechanics Department of Politecnico di Milano, and the friendship and good support of all of the curators of the assorted museums, this program is a real basis for a better future.

We are pleased to have many letters and e-mails from the thousands of visitors to Altino, Ferrara, Ancona, Egnazia, and Sibari, members of the National Archaeological Museums of MiBACT, as well as from the museums themselves, indicating how much the visitors enjoyed the use of the "game."

This program is intended as a gift to our Italian culture, its large repository creating a new micro Renaissance, making the "game" accessible and visible to all, by means of the many forms of communication available today.

The 10,000 Italian museums and the countless tourist centers will become a network for a new small cultural Renaissance, geared towards social learning, with the latest artificial intelligence counted among its creative resources, and applied to the cultural gifts that Italian Museums have.

A community DM can be built. It could serve as psychological and cultural support for people, to increase their knowledge and their feeling of well-being. It can serve as moral support for people in crisis, who feel weak, even for those in analysis, to be honest. It can also serve as social fabric for our times, and help prevent suicides. It can help in the creation of a framework of knowledge, participation and faith that can overcome the destructive tendencies from which terrorism and absolutism are born. DM can strengthen and update information, with contributions to the established knowledge. It can be verified with the appreciation and support of the entire outside world.

DM can fix the points of engagement, indicating the strands through which good deeds can be connected, almost like a sort of religious-secular interface.

In short, it is a good new method for carrying out an important job.

Thanks to all of you, the first witnesses and ambassadors of DYNAMIC MUSEUMS. We thank you in the present for your contribution to the future.

8.2 A Choir for the March Toward Our Goals

What new ideas could be used?

- A short film with the most significant images of the Museums
- A map from Google that zooms out quickly from the seat of the Museum, to show a photograph taken from above
- Short, very simple and dynamic clips, focusing on three objects from each museum, with background music and concise explanations
- A setting for each object selected, based on its time and its environment, with some pictures and very dynamic colors
- A holograph of a single object as an example, to demonstrate the presentation
- Creation of the final image with the aid of logic and emotion, responding to requests submitted by the visitors
- Dynamic explosive effects, which enhance personal expression and which serve as a sign giving voice to the inner emotion of the Visitor, representing it visually, as in the best modern—and ancient—musicals
- Music in the device that accompanies the presentation of the results, storing pleasing results and representing them in new versions.

8.3 One Story

He had always gone to Museums, as a child, in line with all of his classmates and the sound of the teacher talking in his head. It was all beautiful, all there for him to learn about and feel, to ask himself, what is this beauty, this history? It was the curiosity of new worlds.

Something was wrong: in his 5 years of elementary school, including visits to museums featuring memorials to the days of Italian independence, and religious paintings with great colors and textured carvings, ancient objects of history, like the medieval weapons of Edinburgh Castle, he had had the exact same source of culture, the same dates, the same basic information, as everyone else. It was not enough.

Courtesy Collection AR

Child and knowledge

I AM, I WILL BE MY HANDS ARMS EYES LEGS HEART AND VISION KNOWLEDGE EMOTION FOR EDUCATION

The curiosity was linked to his work, and he went to look for documentation on the job, in texts, objects, buildings built to survive through human history.

Never, not once, had anyone asked him: but what would you see? Do you want to bring your curiosity or your sense of beauty, or your strong appetite for the objects of nature and of life, or simply feel the pleasure of beautiful clothes?

If you were today given an opportunity like that furnished by DYNAMIC MUSEUMS FOR SOCIAL LEARNING, would you seek out objects that could be yours, that were suitable to your life? Or would you prefer the story of the creation of systems for generating food or fruit or potatoes? Or … would you like it all together? Would you like it if you, the world visitor, were the center of everything, and not a passive spectator, with everything standing as essentially inert, if not for your movement?

He felt it was possible for him, himself, to be the visitor at the center of Museums, of all the world's Museums. A few years earlier, all the ways of the world had been put into a computer, and you could find your way home from any place, led by tablets or phones, even aided by voices, which would suggest the route you should take!

Thus, why can you not be guided to that which you like, completing your experience, and widening your feelings, and discovering the grandeur of centuries lived and inhabited by humans, everywhere?

He dreamed of writing on a cloud, writing his feelings, his logic, his favorite pleasures, to know them, to live them. Then, you would click, just press a button, and a cascade of images and sounds and music, showing smart, sharp objects, found in museums, would appear. We are divided by space: those who are close, like neighbors, those who would ask you to travel a short distance to see them, and those who want you to reach out to them from much further distances. But are your favorite objects not the statue or the picture or the car that you've already seen, and that you want to go see over and over again, often to confirm your psychological reality?

"DYNAMIC MUSEUMS FOR SOCIAL LEARNING" is the project that will respond immediately to your phone, as you rattle off figures, photographs, information pertaining to so many objects of beauty, kept in Italian museums, ready and willing for you to invest yourself in them. And in addition to this, you can also find all the information you need for travel, hotels, tickets, everything, absolutely everything. And you can print, even with a 3D printer, a copy of the object you intend to visit, before or after seeing it.

In a few years, visitor centers will function for everyone, the wise and the naive, the thrifty and the spendthrift, the quiet and the dynamic, as a conduit to the ways of the world, their guide through the endless streets of knowledge of life and the pleasure of beauty, even attentive to their inner lives.

In less than two thousand days, every object within the Museums representing all the human feelings of every race and country will be made available. We hope to do something beneficial, by proposing the DYNAMIC MUSEUMS FOR SOCIAL LEARNING project, which will hopefully become active within all of Italy's Museums.

8.4 Reconfiguration of Museums with the Latest Scientific Upgrades

Natural science and computer technologies have been invaded by new experiences and new knowledge about the connections between the hemispheres of the brain, illuminating the blend of the logic created by the left side and the emotions interpreted by the right side.

8.5 Reconfiguration of Museums with Upgraded Scientific Knowledge and Communication Technologies

Natural science and computer technologies are being invaded by new experiences and new knowledge about the connections between the hemispheres of the brain, illuminating the blend of the logic created by the left side and the emotions interpreted by the right side. Careful studies are being carried out regarding the results of optogenetics and new applications of fMRI, functional magnetic resonance imaging. The work is centered on knowing and understanding how the neural circuitry of the brain operates, how remote areas can communicate and how the almost one hundred billion neurons and their connections can give rise to the mind, creative thinking, moods, behavior and emotions. The world, furrowed by the connections between information and culture, communication and participation, is enriched in its logic and its emotionality. Everyone can now be present as themselves in every place and at every time.

The technologies of the mind are still comparatively weak and fragile: the degeneration of the mind that we continue to see, in the young and elderly, in new, unexpected numbers and varieties, destroys any easy optimism we may wish to maintain. Today, we still do not know how to deal adequately with Alzheimer's and Parkinson's. The age of people broke through the wall of maturity, and we are required to understand interpretations of life for everyone. Neuropsychology is immersed in the environment and in human situations, with courage and determination, for a broader interpretation of emotions. Connectivity is the new basis for each study and all knowledge. **The brain maps of the first neurologists in 1990 who developed the flight of supersonic aircraft with command brain, appear in cultural archeology**. The inseparability of the two cerebral hemispheres has been reassessed in recent years. Knowledge is now heavily reliant on the Internet and multimedia communication and computing. Data memory and facts are stored within clouds and electronic devices with great capacity to handle any amount of data. Hidden wars are taking place between powerful groups with the goal of managing the world's knowledge and information, the basis of any economy. All knowledge entails secretiveness, political, military realities of power and defense, and security needs of the future. Cultural movements mark the paths that are not only rational, but often highly emotional, to love things and to stay far away from worries and fears. The visual patterns, communication over the Internet that is in full color, with music or sound, recall emotional and physical participation.

The sharing of information, knowledge and learning, which becomes more difficult and demanding, is also reinforced. Infinite data is managed in the cloud, like a department store carrying the power of all knowledge that is remotely accessible in mere nanoseconds. Data policy is linked to the intelligent management of the flood of information, and new algorithms are being developed that know how to manage logic and emotion with a wisdom that could almost be seen as inhuman, if they were not the result of humans.

Among the symbols that interpret modern humanity (and its vision towards an auspicious and positive future), knowledge is represented by museum collections. They are expressions of the culture that encompasses all of our thoughts, all of our actions, even those deepest within us, as well as our hopes and dreams.

Courtesy Museo Marinaro di Camogli

Sea and work

A SAILOR IN A VILLAGE IN THE HARBOUR, WITH FAITH AND HOPE IN THE SEA AND IN ACTIVE LIFE

8.6 Redesign of the Museum Path

Knowledge over the centuries was transmitted through books and stored in libraries. From the birth of the digital, knowledge has been stored in powerful digital systems, with limited space and subject to fast and easy access, from anywhere in the world. The aspect of culture expressed through significant objects such as paintings, statues and archives are preserved in museums. In recent times, visitation to museums has grown by high percentages. Museums represent the expression of knowledge: we consider museums to be the first protagonists of a new acquaintance with knowledge, the sort that computer networks cannot provide by themselves. Museums are a domain of great importance, which is growing day by day, as users increase in every social and economic sector. They are dedicated, in the most traditional sense, to the collection of objects. Today, however, they require intelligent objects, which must be visible, readable, understandable, acceptable, and possibly interactive with the visitor's pleasure and emotion. From now on, museums will properly be designed to be "smart," that is, each "object" will be suitable for gaining understanding through logical and emotional participation. The human motivation to visit a museum is linked to a desire for knowledge, empathy, and even emotional limits with the potential to create very dangerous effects, referred to as the "Stendhal Syndrome," which leaves an unpleasant sense of inadequacy. The most significant acceptability is linked to personal passion, the satisfaction, the excitement of standing in front of quiet beauty, curiosity satisfied, the emotional memory fed. The two hemispheres of the brain are in a continuous collaboration and empathic communion during a visit to a museum.

Museum collections consist of donations, series of objects and works, themed sets, personal experiences and historical events, related to History, the Arts, Industry, Science, Fiction, Nature, and many other subjects.

The organization of Museums is very variable: physical paths featuring objects with demo panels of physical descriptions, mixed systems, virtual tours, itineraries along digital paths involving multimedia, knowledge services and selective displays, customized, and digital communication.

However, much—too much—of Italian culture and that of the Italian Museum is unknown, to both foreigners and Italians alike. The ambitious goal of the DYNAMIC MUSEUMS FOR SOCIAL LEARNING Project is to increase income from tourism by creating a form of visitor service tied to museum fruition that is sustainable and can be applied at various levels.

Basically, the DYNAMIC MUSEUMS FOR SOCIAL LEARNING Project would offer a New Cultural Service.

The DYNAMIC MUSEUMS FOR SOCIAL LEARNING Project decided to take a new road that would suggest to museums throughout Italy ways to improve their relationship with their visitors. Through talking to many different tourists, both Italians and foreigners, those educated or simply interested in beauty, art or technique, we can suggest the objects and the routes, the information and the knowledge that can make it even more exciting to visit a museum. Digital technology is an integral element, allowing us to share information, thoughts and feelings. PCs, tablets and smartphones, connected to the network, allow for the considerable expansion of the catchment area of the DYNAMIC MUSEUMS FOR SOCIAL LEARNING Project application.

8.7 Possible Results

The DYNAMIC MUSEUMS FOR SOCIAL LEARNING Project has focused its energies on building a futuristic reality whose roots are the results of today and whose branches represent the results of tomorrow. The force of modern knowledge can overcome all obstacles of time, distance, environment and culture, if well guided and armed with precise objectives. In short, it pulls together all knowledge of computer design, communication design, data management, virtual and multimedia representations, and multivariate mathematics to show how each person can be brought more and more into the reality of the museum. This not only considers the predetermined path and the programmed set, but also the free choice of knowledge and participation, born out of intelligence and passion. In this way, the visitors find themselves as the new front of the intelligent object, the work of art, the archaeological piece, the scientific instrument. The quality of the items (as Aristotle wrote in 340 BC) attracts human attention, in its logic, in its emotional and vital participation. It is a contribution parallel to the imposed and programmed path, which remains effective for a priori suggested knowledge.

8.8 Evolution of the Museum Visit

The DYNAMIC MUSEUMS FOR SOCIAL LEARNING Project has prepared an evolution for the visit to the museum, which is becoming pervasive and global. The development is based on the use of new knowledge about psychological neurology, which has been confirmed by the scientific community, but which gained little attention until a few years ago.

The "DYNAMIC MUSEUMS FOR SOCIAL LEARNING" Program starts from the knowledge of the activities of the two hemispheres of the brain, the left one for logic and knowledge, the right one for the elaboration of perceptions leading to feelings, emotions, emotional involvement. For the past few years, activities have been designed and developed with new knowledge of the workings of the right hemisphere.

Should not the same apply to museums? The reality of the significant richness of Italian culture? Why not turn the new knowledge of our intelligent nature into a cost-effective approach to our national reality in Italy? All knowledge in communication design, image management, economics of information, the use of multimedia techniques and virtual reality, artificial intelligence, and multivariate mathematics, so far used mainly in technological and scientific projects, can be and has been applied here to this end. The designers focused their eyes and hearts on intelligent objects located and often hidden within our museums, large ones in big cities, even small ones in unknown villages, those of the sea and of the mountains and of the plains, with a method that is both "aggressive" and pleasant.

Courtesy Museo Marinaro di Camogli

Blue image

BLUE SENSATION OF FREEDOM AND POWER

The pleasantness of the method comes from the fact that it asks each visitor, as an individual person, to read a simple table depicting logical activities and emotional perceptions, which also constitute the main activities of the individual's life. The tables seek to spur an interest in logical evaluation and emotional involvement on a scale from zero to ten. The numbers can be entered into their mobile phones, in a simple colorful table, which is transmitted to the DYNAMIC MUSEUMS FOR SOCIAL LEARNING program control computer. The table, for the sake of privacy, asks only age, sex, occupation, country, and, if the visitor is willing, an abbreviated nickname. The Visitor table is prepared so as to be able to be considered by all the Museums in the network of the DYNAMIC MUSEUMS Program.

8.9 Input for Transforming Stored Materials into Exhibited Materials

In addition to the catalogs available in the museum itself, the choice of information about museum sites can be gathered on the Internet, through search engines like Google, Baidu, and others. It is vital to have a link to the museum and a selection featuring all linkable museums. The quality of information brings up issues of value and significance. The ranking of the information can be according to criteria of truth, quality, completeness, presentation and even the merging of text, images, video and links. Even the organization of classification is essential. For objects collected in the Museum, there is always ongoing news, which requires monitoring and verification, accompanied by a projection of what significance the object may have in the future.

In the museum, the protagonist is the person, with his/her conscience, with her relation to the outside.

Today, museums can follow the evolution of communication, presentation, and shared participation. This relies on the left hemisphere for logic and thinking, and the right hemisphere for emotion and action. Knowledge and modes of thinking and living have been caught up in a whirlwind of change over the past decade, as the world becomes increasingly stretched and relaxed from the past to the future.

The curator of a museum prepares a selection of significant objects, divided into groups through use of a simple file, lightweight and easy to manage, containing images of each item proposed. The same table is compiled by the museum curator (or an expert/delegate) indicating the logical importance and emotional interest that the Trustee believes may be transmitted through each intelligent object on display, with values from zero and ten. As many data packets as the curator likes can be prepared.

Once ready, the control center computer intersects with multivariate mathematics and neural network systems, sharing data and information on each item suggested by the curator. In a matter of milliseconds, thousands of museums can be explored, and the visitor receives, in return, an ordered and precise classification of objects which will fulfill him and give him the pleasure of knowledge and emotion, selected from all connected museums. Thus, in a few seconds, the network provides the visitor, who may be located thousands of kilometers away, with a ranking of figures of favorite items, which have been selected and proposed for him/her and him/her alone. This is a gift of technology for his/her ego, his/her knowledge and his/her pleasure.

The beauty of the result emerges from a table, created by the network computer, indicating which museum objects correspond to the person's interests, and indicate the location of the Museum, the dates and hours for visiting, the geographical position of nearby restaurants, means of transport, and all other facilities required. The system provides an output package of more addresses, suggested per the requests of each person. This transverse dimension of museum visits offers a significant deepening of the museum heritage, which is now no longer simply consolidated for decades or centuries of appreciation, but also for a singular richness of life and culture in which new actors and creators offer new products, for greater cultural richness.

8.10 Input for New Designs and Realizations of Museums

The same network updates the "container" that the person represents in an automatic and conscious way with icons, through multimedia means. **Everyone is looking towards tomorrow, to act in a life that increasingly calls for presence**

and participation. That requires consciousness of being open to knowledge from the outside, in respect to the construction of a planet increasingly rich and purified of darker forces that operate without logic and without positive emotion.

The main goal of the Project is to create a network of all interested Museums in Italy, making the subsequent database available through the Ministry of Heritage and Cultural Activities and Tourism. The Ministry's expertise in defining the importance of each museum can be used for developing the software program. A search of the museums would take the historical importance of each Museum and each collection into account.

8.11 Design of the Museum

One great positive effect of the "DYNAMIC MUSEUMS FOR SOCIAL LEARNING" Program is the possibility for the DESIGNER OF THE MUSEUM to modify and update the museum presentation based on the requests and responses of the visitors. The designer has a perfect source of information from the analysis of visits and comments from visitors, all the more handy when collected on the network. The museum unit visited is only the starting point of a virtuous circle that delights the visitor, who may go on to seek out natural environments, new cities, or new countries, completing his/her cultural identity in a beautiful way of life. The "DYNAMIC MUSEUMS FOR SOCIAL LEARNING" Program asks the visitor whether they prefer to visit towns or cities, whether they would like to find museums in unknown locations or if it doesn't matter, what the priority given to his/her intellectual and emotional curiosity should be. Through calculation, the program will change the rankings of selected objects per these directions, providing a more enjoyable ranking for the visitor.

The first consideration that arises is related to the advantage offered by the "DYNAMIC MUSEUMS FOR SOCIAL LEARNING" method. First, human beings, being able to read and write (although, in the future, through the use of simple icons, it may not even be necessary to know how to read and write to participate) can fill the logical and emotional interest tables on a handheld device or tablet or computer, as well as include certain data pertaining to sex, country, occupation, and main fields of museum interest, if desired. After a few seconds, the DYNAMIC MUSEUMS system suggests objects from museums that match the interest of the visitor. The visitor has now joined the ranks of possible users of those indicated museums, and if the presentation is appealing and attractive, the virtual and remote user can decide to visit the museum and/or enter the museum or particular objects therein into their tour route. Of course, the system also provides travel distance and possible points of accommodation, food and support for the visit in the chosen venue. It is quite easy to assess the economic value induced by this virtuous circuit, both for the structures welcoming the visitors and for the visitors themselves, who get the satisfaction of receiving what they have requested. The visitors build and create their own personal paths. In addition, the local restaurants and methods of transport mobilize the resources that bring further attention to the Museum and its riches.

With the "DYNAMIC MUSEUMS FOR SOCIAL LEARNING" Program, the users can see all the suggested items and choose them, or settle for a display on their mobile screens and computers, reading all related information. It is in the interest of the Museum to provide the best documentation possible, so as to acquire a new customer. The object can also be presented through holograms, models made with a 3D printer, videos, files, and multimedia communications, at the lowest possible cost.

In conclusion, the objective of the DYNAMIC MUSEUMS Project is to offer a service to Italian culture, with an amplification of the management of Italian Museums, even those lesser known, for the maximum exploitation of the cultural heritage. The program opens new horizons to people who are not experts in museum heritage, and therefore is a lever for increasing tourism, defined as "cultural and museum tourism." Also, it certainly improves customer satisfaction, i.e., the approval of the visitor, since they are offered only those objects that matter most to their culture and emotional perception. They can obtain information on their favorite museum objects at no cost, except for the small fee for participation in the app. Moreover, the curator of the museum, by the characteristics of the new visitor, can then evaluate a redesign of the museum, as is happening with the museums that are already participating.

8.12 Panorama on the "DYNAMIC MUSEUMS FOR SOCIAL LEARNING" Project with Emotions

8.13 Make the Visitor a Player

The elements that constitute the overview are:

- Vision of the DYNAMIC MUSEUMS FOR SOCIAL LEARNING Project
- Mission of the DYNAMIC MUSEUMS FOR SOCIAL LEARNING Project
- Main Features of the DYNAMIC MUSEUMS FOR SOCIAL LEARNING Project Partners
- Phases of the Project
- The DYNAMIC MUSEUMS FOR SOCIAL LEARNING Project and MiBACT: synergy, advantages and results
- Next Steps.

The Vision of the project consists of:

- Make Italian museums and the Beauties of Italian Tourism easily accessible
- Turn the visitor into a player, as opposed to a mere spectator, in cultural activities and tourism
- Transpose the feedback and opinions of the visitor
- Determine the visitor's emotions and desire to participate, including in deeper forms of learning.

The Mission of the project is divided into three parts:

- The computer tool created, that is, the "Easy Visit" program, is an appealing platform that attracts and personally involves the Visitor
- Use of the new knowledge of neurobiology to promote the culture and the quality of people's lives
- Individualization and customization of the development paths of museum culture.

The comparison and collaboration with large institutions will make the "DYNAMIC MUSEUMS FOR SOCIAL LEARNING" Project widespread and viral.

The Main Features of the Project are:

- The functional simplicity of the Program that fully involves the Visitor
- Results catering to the Visitor's expectations through the suggestion of custom locations
- Recent developments in artificial intelligence involving deep learning, in addition to mining processes.

The table represents the 16 items specified by the National Archaeological Museum of Ferrara.

The program, developed with the Convention with the Ministry of Heritage, Cultural Activities and Tourism, has been utilized in 5 National Archaeological

Museums, those of Altino, Ancona, Egnazia, Ferrara, and Sibari. This program is specifically designed to cover National Archaeological Museums.

Phase I of the project: the value acquisition

This image is displayed when the program starts: The Visitor, through use of the touch screen, expresses the desired value of the 8 parameters of each "diamond." The choice of the meanings and values depends on the moment and the logical-emotional state of the Visitor.

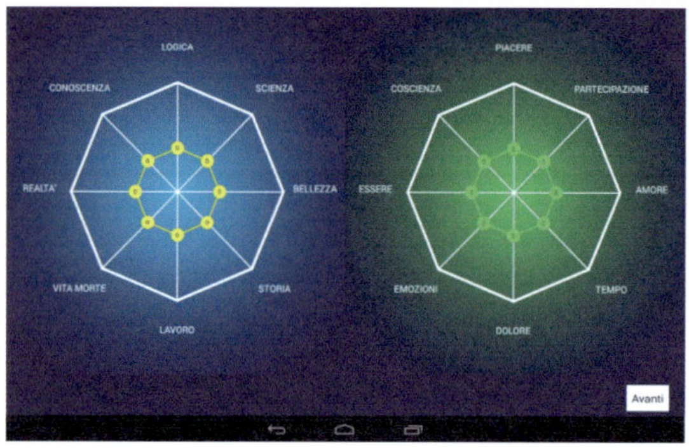

Courtesy Collection AR

Phase II of the project: characterization of the mood of the visitor

The figure indicates the values—from 0 to 10—chosen by the users.

The Visitor then takes note of the signs representing the two hemispheres of the brain that underlie behavior.

Here, the Visitor's values begin their intersection with the values shown by the Museum Curator for each object.

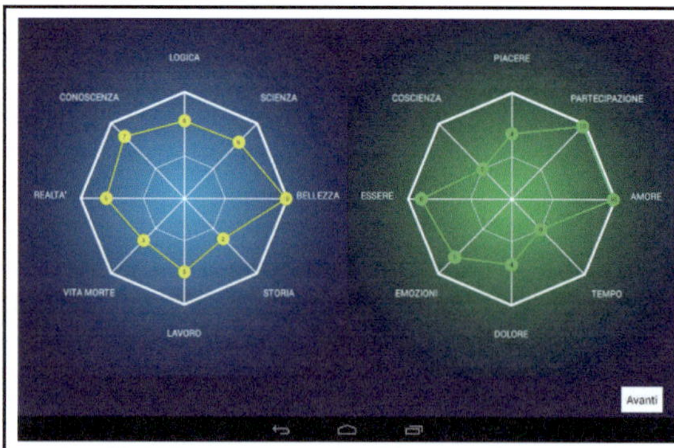

Courtesy Collection AR

In the figure, the path generated for a guest of the Museum of Altino is shown; on the upper right, the preferential score for each object is shown. The indications of the Visitor and the numerical notes created as a response represent the three “smart objects” in the figure, the first among the 16 objects selected from the Museum.

Courtesy Collection AR

Phase III of the project: development outcomes

- The development was carried out in collaboration with the five National Archaeological Museums selected by MiBACT, the Ministry of National Heritage, Cultural Activities and Tourism
- More than 1300 people have used the Program
- Positive Rating: average 8.2 out of 10
- There is interest out of MiBACT in expanding the pilot project to a network of National Museums.

Phase IV of the project: benefits for the user

- Immediate expression of the Visitor’s state of mind through logical and emotional indices
- Emotion of the Visitor in regard to the results, because they are carefully customized to logical and emotional preferences
- Loyalty to the museum and intense participation in the visit, as an expansion of one’s personality
- “Love” for the results, expressed by the Visitors in an 8.2 score out of 10.

The DYNAMIC MUSEUMS FOR SOCIAL LEARNING Project and MiBACT

- The DYNAMIC MUSEUMS FOR SOCIAL LEARNING Project creates a path for the Visitor/User based on stated preferences and deduced state of mind
- The DYNAMIC MUSEUMS FOR SOCIAL LEARNING Project allows for the collection of data on Visitors/Users in significant proportions (logical and emotional)

– The DYNAMIC MUSEUMS FOR SOCIAL LEARNING Project contributes to a new culture that expands both the wealth of the Italian Museums and their exploitation.

MiBACT caters to 440 curated museums: the DYNAMIC MUSEUMS FOR SOCIAL LEARNING Project is a potentially useful tool, because it adopts methods for knowledge management through logical and emotional patterns, along with criteria derived from Aristotle (Works, Categories).

The collaboration with MiBACT can make the application more accessible and usable in the worldwide network.

The advantages for MiBACT

- User information and data combinations allow for the customization of target profiles
- Enrichment of user profiles
- Customer Acquisition
- Growth of the number of Museums connected through MIBACT and the DYNAMIC MUSEUMS FOR SOCIAL LEARNING Project
- Opportunities for immediate use of the DYNAMIC MUSEUMS FOR SOCIAL LEARNING Project in cooperation with MiBACT to make the Visitor/User's cultural experience more participatory, more personal and more emotionally engaging
- The chance to offer, through MiBACT and the Easy Visit App, a choice of routes in Museums among those suggested, and the ability to select events and cultural activities that will increase the number of users of Italy's heritage and wonders.

Product configuration and transmission of results

- The product configuration will be studied after acquisition of the MiBACT model, through the deepening of mutual information on the platforms and based on the primary objectives that MiBACT holds
- The results are transmitted via the DYNAMIC MUSEUMS FOR SOCIAL LEARNING Project, with patent pending and machine learning methodologies that MiBACT can use.

Next steps

- Sharing the Project
- Project Rating
- Extended application throughout the Italian territory
- Worldwide communication with Visitors and Tourists, encouraging them to visit Italy
- Documentation of events, theater, cinema, exhibitions, fairs, etc., also for touristic enhancement.

8.14 What These Pages Deal with

These pages describe the birth, growth, and development of a project dedicated to museums, and consequently to the culture, with respect to the application of the most modern communication technology, computer science, neurobiology, and artificial intelligence. Today’s technologies of communication and information have involved the world of culture, almost to the point of overwhelming it. Starting in an unconventional way through concepts of intelligent robotics, these pages show how we can now open up new cultural horizons. communicated through the computer, the mobile phone, the tablet and all the most appealing computer means, creating and nurturing cultural knowledge. This project applies the criteria of the new artificial intelligence, reborn and generated since 2008. It is founded not only on algorithms, but also on the careful acquisition and use of learning concepts that, in absorbing very large databases of information, can apply intelligence, even in a way that introduces emotions into the logical process. A new world is really growing before our eyes. These pages tell how this pertains to the new project called “DYNAMIC MUSEUMS FOR SOCIAL LEARNING,” because it elaborates on our knowledge of logic and personal feelings towards objects of the past, which today are called “smart things” (Intelligent Things), to make us even more active participants in our current life. The beauty of life subsequently explodes within the culture, having once been the domain of only a few. The person becomes the center of the culture, a true character, and not a passive and commanded Guest. The person can see into the spaces of all the museums of the world.

8.15 One Story Is Born

This story comes from the thoughts, words, and actions from people’s lives. The events are intertwined. The “DYNAMIC MUSEUMS FOR SOCIAL LEARNING” Project has developed a software program that puts the visitor at the center of the museum visit in a new way. No more buildings to go through in sequence per preprogrammed tables, but rather a new approach to the use of museums. This means providing tools so that every person can be a participant, a protagonist and a creator of their own visit to Museums. Visits that are personal, profound and still very much their own.

8.16 Intelligent Agents

The strength of the DYNAMIC MUSEUMS FOR SOCIAL LEARNING Project is in its ability to recreate the reality of things, which appear in their multifaceted nature, today referred to as “intelligent agents.” The Curator of the Museum colors the objects that are considered important and beautiful in the museum with the colors of logic and emotion, thought and enthusiasm. The desires of the Visitor intersect with the advice of the Curator, in a human algorithm.

8.17 Human Conditions

These written words tell of the today human condition that desires to remember the things of the past, keep watch on the things of the present, and anticipate the things of the future, many of which may be virtual, born from the imagination, from desire and hope. The computer program "DYNAMIC MUSEUMS FOR SOCIAL LEARNING" represents a significant track to follow into this new world, where many models of knowledge and types of knowledge can be communicated, offering greater and greater levels of beauty. This book wants to explain to you how you can now build a personal visit to the museum, showing the means to partake of its knowledge. This ability can increase our knowledge and our participation in the profundity of cultural life, with the help of "smart" things, i.e., the Museum objects around us. We also integrate the natural beauty of the environment into the cultural offerings of our cities and territories. We make our own wealth, consciously and emotionally. We are also grateful for the new knowledge about the human brain, biogenetics, neuropsychology, and neurobiology. They show us how our minds/our souls have infinite colors. A tomographic photograph of the brain teaches us that culture is always associated with emotional participation. In front of a beautiful picture, a beautiful landscape, emotion is important. We enjoy our growth over the same years, cleansing the brain of the clutter of the past—neurobiology as understood after 2000—forming building blocks for new sensations in our daily existence. As we learned a few years ago, today, neurobiology is a part of daily life.

8.18 The Gray Wire

The gray thread that unites the entire book.

The book may have different colors, one for each part. The first and the second parts are dedicated to the description of the scientific and economic criteria that set the terms for the operation. The three subsequent parts are devoted to innovative criteria within the communicational sciences, which make the DYNAMIC MUSEUMS FOR SOCIAL LEARNING Project unique. The last part is devoted to the results obtained from the Project at the 5 Museums of the National Archaeological Park, from thousands of documented experiments, with the sponsorship of the Milan Politecnico Foundation and MiBACT.

8.19 Towards My Italy

First and Second Parts

8.19.1 Planning

The first and second parts are devoted to the operational setting, and the cultural and technical management of the DYNAMIC MUSEUMS Project in relation to the Italian State Museums and the Ministry of Cultural Heritage MiBACT in Italy. After an introduction to the history of museums, as experienced by a child, the software for the project, capable of handling the emotional character and psychological complexity of a person, is explained. In this way, the museum is reconfigured, because the visitor seeks to know the inside of the museum with the help of the Program, following the right path for their character and their sense of fulfillment. Consider that the program can be applied to all museums, cultural works, monuments, and even meetings, shows, and events taking place in the city, the region, the nation, the world today, because use of museum data and big data management through modern computers is a very big task, but also a simple one.

8.19.2 Valorization

The project design needs to take clear inspiration from the Visitor who simply wants to be inside the museum through the immaterial yet profound participation of the Curator. The choice of objects that are more pleasing to the visitor requires the use of concepts of modern neurobiology. The project especially applies to the interweaving of emotions with logical criteria that, until a few years ago, were considered essential only for choices of behavior. The contents of each object chosen for the Visitor need indications of knowledge and emotional enjoyment. The result is a policy for the promotion of the objects that speaks particularly to the intimate behavior of the Visitor. The results of the “DYNAMIC MUSEUMS FOR SOCIAL LEARNING” Project produce a platform out of the data given by each Visitor to create reference models that can help in the redesign of Museums.

It also builds a range of information that grows, creating a sort of archive out of the broad participation of the Visitors who are willing to offer up information about themselves. It is a unique database that becomes an extension of the Visitor.

Courtesy Collection AR

8.19.3 Platform

The project consists of a very wide platform that receives information such as basic data about the Visitor. The "objects" are the reference points and evaluation by the Museum Curator. All results are crossed with logistical and environmental information, related to events and territorial aspects of tourism. Initially, the visitor shows up and declares, if he/she wishes to, sex, age, occupation, fields of study, etc. The visitor approaches the visit and is presented as an individual, as a couple, as a family, as a school group, or whatever. The software is prepared for many real-world conditions. The platform is an operational scientific virtual object for communication and information, and as an application, it is managed by the Project Science Center. The application is downloaded from the same platform per the users, following economic opportunity management, for the payments of regulations and commercial structuring. In addition, all data can be used to build a function for the life of the user, who applies the information gained from the historical memory.

8.19.4 Economics

Economic aspects and employment aspects are important. Instead of conceiving of a program out of general concepts, we preferred to indicate in this book the Italian reality in all its artistic beauty, spread across the territory of the Museums' reality. You can also create new working conditions with a fresh focus on full harmonization and organization of the museum experience as a spark of participatory knowledge for the development of these national resources.

8.20 Brain, Mind, Soul

Third, Fourth, and Fifth Parts.

The third part concerns the intellectual, scientific developments regarding neurobiology today. We start on the basis of behavior, an essential union of the two cerebral hemispheres. We want to connect the important things that are now called "intelligent" with the person, who, in the "Dynamic Museum," reflects the past into the present with a deep sense of today's self.

The fourth part deals with features of dynamic museums for social learning in action.

The fifth part presents scientific and cultural trends.

8.21 A Project, A Bridge

Sixth Part

The project proposed to the Politecnico di Milano Foundation was presented as an important development for Milan and Italy. It was a bridge, which has been proposed by the Fondazione Politecnico to MiBACT, the Ministry of Heritage, Cultural Activities and Tourism in Italy, and it was subsequently stretched to the Convention. It took many busy months of development at five Italian archaeological museums starting on August 2nd, 2016. The museums are those of Altino, Ferrara, Ravenna, Egnazia, and Sibari, very well chosen by the Ministry, as the five are spread across Italy. The results were the acquisition of information from more than 1600 people who used the DYNAMIC MUSEUMS FOR SOCIAL LEARNING Program in their visit. The program runs on a platform allocated within a tablet that is available to the Visitor and on which the Visitor indicates their logical and emotional preferences. Back in July of 2016, at the World Congress of Museums (ICOM), the project was invited abruptly to take part in two presentations. The applications were a hit with people of every nation, color, language, and culture, all in the first week of July 2016. On August 2nd, 2016, development began in Italy at the five national archaeological museums listed above, with the immediate help of all of the museums' curators, fascinated by the technological and cultural innovations, which were both unexpected and very human. It reaped the results, which were compiled into a dossier, to be submitted to the Commission of the Italian Senate. Meanwhile, institutions, agencies, companies, industries, small towns, small museums, and major museums are all being contacted so as to gain experience, collect information, and even costumes, for a new cultural world expressed through new technologies, to help us deal with the excruciating reality of today.

Part III
Cultural, Vital, Emotional Approaches Within Museums

Chapter 9
Leading Indications for Cultural Notes Through "Dynamic Museums"

9.1 Towards the Museums

1. Towards the small museum, local, almost isolated, and because it is in the network, available and accessible, with the possibility of being known in greater depth, linking every object with the database of the Museum towards DM
2. The Small museum is rewarded, because it is accepted, visited, and appreciated, with its indexes being entered into the network in the DM-circuit repeatedly
3. The Small museum is appreciated by visitors, and therefore it is raised from the informational level to the cognitive level
4. The results of the visit of each visitor is saved by the DM program, and are linked to the Museum, if you like, to serve as a reference flag for the visitor doing research on the net
5. It creates a growing list of sites/favorite items, such as Golden Objects, with a photograph, a documentation link (short, medium, long) and the possibility of a reproduction created on a 3D printer dispatched from Amazon or another company
6. "Drops of wisdom" are created and saved in various languages, which allows the number of items to grow
7. You must create a beautiful pattern for the platform, easy to look at, with colors and familiar music, so that it can become a must for those who love museums, those who are equipped with badges, special tickets, and an interest in engaging in MODERN COMMUNICATION
8. Creating a useful MINI NETWORK representing the geographical locations of Museums, objects of interest to tourists, streets, hotels, etc., and represent it with Helvetica (the font used by metros around the world, because it is clear and distinguishable)
9. Creation of a useful TELEMATIC THEMATIC NETWORK, with a total digital offering for the visitor based on their demands and conditions.

A. Rovetta and E. Rovida, *Scientific Knowledge Communication in Museums*, Research for Development, https://doi.org/10.1007/978-3-319-68330-0_9

Courtesy National Archaeological Museum of Marche
Travel

9.2 Visitors to the Different Models

1. Local Guest, who lives in the area and wants to deepen their understanding, see things, discover things, review things, love things, etc.
2. Regional Guest, close enough to the Museum to get there within a few hours of travel time, able to perceive, understand and love the culture of the Museum, and who can discover what DM can give to them: photos, 3D model, holography, links, figures, etc., plus opinions of other visitors from all over the world, and not just the positive opinions
3. World Visitor, from infinite possible locations, even overseas, whose visit will be enhanced by the sort of magnificent planning that, WITHOUT DM, would otherwise be lost
4. The Visitor as a single individual
5. The Visitor as couple, and therefore compatible with the couple profile on DM
6. The Visitor as part of a group
7. The Visitor as a member of a formal and exclusive, even snobbish, club
8. The Group of visitors of various types, and of various cultures, without a common base
9. Advantages of the Visitor to the Museum who uses DM compared to those who go it alone: simplicity, a wide range of response to one's own curiosity, suggested routes, including those recommended for the total tourist experience, option to engage with artificial intelligence after selection of the path; recommendation of the path most likely to be loved by the Group, the club, the team, the Agency, etc.
10. Collecting the views of Visitors and choosing, through the optional programs of DM, the most consistent evaluations
11. A beautiful spectrum of nuanced answers to the Visitor's questions through DM's configuration programs, with vibrant colors and good music,

participation that will make it that much better for the subsequent Visitors, adding to DM's already substantial value.

Courtesy National Archaeological Museum of Marche

Sarcophag

9.3 Psychological Aspects Through Dynamic Museums

1. Consolidation of the self, even at advanced ages
2. Enrichment of the self through new knowledge and a new vision of life, (e.g., some works of Michelangelo, Kandinsky, etc.)
3. Curiosity, often unexpected and surprising, but marked by a lack of participation until Dynamic Museums arrived
4. Retention of the visit as a fond memory, remembering the restaurant you visited, and the book shop, and becoming a member of the news network, simple and appealing

Courtesy National Archaeological Museum of Marche

Glass

9.4 Creating Corporate Start-Ups, by Applying Expansion of DM to Museums and Tourism

1. Undertakings in tourism, such as innovations and start-ups, did not exceed a turnover of 18 million in 2016; this is a very low value, indicating the current lack of innovation in museum/tourism relationships; DM can quickly address this problem, as it has for many other areas where the network rate has changed. Therefore, DM can create Start-Ups for:

2. Plotting out the best Museum paths
3. Supporting the choice of Museum visits within tourism
4. Preparing appealing documentation of the Museum paths

Courtesy National Archaeological Museum of Marche
Decoration

9.5 In Regard to Monetary Implications

1. Funds from Visitors for membership in CLUB DM, collecting all the bookmarked data, with high-level selection
2. Creating a presence for the Visitor in a preferential network of Museums, with automatic enrollment in the network Club, through which they will receive exclusive information, select invitations, discounts to events, etc.
3. Exchange of suggestions guided by artificial intelligence with appropriate learning programs linked to the character of the Shareholders
4. Virtual Meetings through holography, 3D objects for participation and/or ownership, and even conferences linked to local events
5. Conferences using holography
6. Super conferences with other high-level network members related to tourism and culture

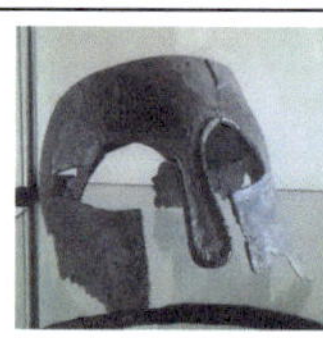

Courtesy National Archaeological Museum of Marche
Protection

9.6 In Regards to Tourism

1. Increase in earnings from tourist visits to museums, both local, on the smaller scale, and global, on the larger

2. Use of the network of museums with a password, and cultural events linked to tourism
3. Network for planning long visits connected to other important avenues of tourism that can suggest other worthwhile visits within the territory, fostering the integrating effect and global communication
4. Preparation of a virtual network, created through the results gathered by DM, with the choice of the most suitable and most pleasant maps for short and long routes
5. Short paths, edited down from longer ones, that focus on the items specifically favored by the user, the overall program being cut ad hoc, with added folk tourism, etc., according to the visitor's individual preferences
6. Creation of individual personnel programs, with each feature and welcome aided by artificial intelligence programs that not only indicate the routes, but also local gastronomic attractions, sightseeing, sports, etc.
7. Strong setting linked to sport and sports activities, with opportunities to be either an active participant or a spectator. This can create a symbiotic relationship in which the sports program and the cultural program come to rely upon each other.
8. Saving the information gathered on the programs used, for preparation of a book that collects the colorful and powerful experiences of visitors, which will also be made available online, with additional information not always present in the official iconography and traditional paths for Visitors

Courtesy National Archaeological Museum of Marche
Wife memory

9.7 Invention

1. New products featuring objects created through digital holography, for a fee, 3D models of which can also be delivered to the individual's home at low cost
2. Circuit series of Museum objects with relevant appeal

3. Creating a "temple" of beautiful things, such as a collection of three-dimensional figures based on the pleasure and excitement of the Visitor, said "temple" also being used to store music, other artworks, etc., built with artificial intelligence models capable of creating the most linear design
4. Systems of automatic translation and links to museums with original aspects, starting with the Museums already involved in the Start-Up, pre-financed
5. Creating ideas for start-up development using a high level of creative design, culture, 3D printers, holography, appealing images
6. Creating start-ups for the purpose of developing specific aspects, to be selected from those of maximum utility
7. Creating courses for other visitors networked within Dynamic Museums, such as the elderly, youth groups, children, science clubs, etc.
8. Drawing up a program that allows visitors to physically build a museum with a laboratory, through both virtual and augmented reality
9. Building assembled systems of museum objects that tell a story
10. Working with what already exists to develop a start-up out of minimal invention, which can be opened up to financially valuable applications within the territory
11. All of this represents the yield of two years' experience within the territory of the Italian museum, through engagement with all of its tours and applications

Courtesy National Archaeological Museum of Marche

Bear skeleton

9.8 Costs of Dynamic Museums

9.8.1 Cost of Software

a. Block diagram of the program
b. Genesis program
c. Program test
d. Program application
e. Ads for the program
f. Program instructions

g. Loading of the program platform
h. Application to Museums and Visitors
i. Construction of the app

Courtesy National Archaeological Museum of Marche

Boxing and sport

9.8.2 Cost of Hardware

a. Hardware purchase
b. Device rental and terminals
c. Placement within museums
d. Placing panels, terminals, etc.

9.8.3 Cost of Personnel

a. Assistants to the Curators
b. Support for DM Museums
c. Secretarial and developmental expenses
d. Legal and tax assistance
e. Economic assistance
f. Secretarial management

9.8.4 Start-Up Costs

a. Contact with institutions (e.g., banks), museums, ministries, etc.
b. Developmental meetings
c. Presentation of cooperation and funding requests
d. Costs for the development of applications in Museums

9.8.5 Costs for Insertion in the Program of Each Museum Objects (Up to 8 Objects)

a. Purchase of photos, drawings, data, etc.
b. Preparation tables and assessments
c. Program development for the Museum
d. Developmental test
e. Advertising and communication network

9.8.6 Cost of Tables of Museum Objects

a. Purchase of photos, drawings, data, etc.
b. Preparation tables and assessments
c. Program development for the Museum
d. Developmental test
e. Advertising and communication network

9.8.7 Communication Costs

a. Illustrations in the press and media of the DM project
b. Distribution of files to outlets
c. Advertising in different languages
d. Distribution in Schools and Universities
e. Support from banks and institutions for advertising

9.8.8 Cost of World Integration

a. Global information
b. Information through the World ICOM
c. Information provided by the individual museums

9.8.9 Associated Costs

a. Corporate expenses
b. Tax and investments

9.8.10 Costs of Innovation and Ideas

a. Development of ideas through start-ups
b. Land development
c. Possible links with both foreign and Italian companies
d. Results of communication costs

9.8.11 Duty and Travel Costs, etc.

a. Transfers between Museums in Italy
b. Contacts and congresses
c. Development of operational partnerships

Courtesy National Archaeological Museum of Marche
Treasure

9.8.12 Cost of Collaboration with Cooperating Institutions

a. Start-ups and collaborations with start-ups
b. Association with consultants and technical experts
c. Collaboration and organization of cooperation

Costs of	1 product	1 year	Yearly
	K EURO	K EURO	K EURO
SOFTWARE		120	
HARDWARE		60	
PERSONNEL		75	
START		120	
MUSEUM 8		2	
MUSEUM 800		5	
COMMUNICATION		110	
INTEGRATION		28	
ASSOCIATION		25	
INNOVATION		40	
TRAVEL		150	
COLLABORATION		45	
		780	

9.9 Dynamic Museums Towards 2020

In 2010, with the Shanghai World Expo, we passed the threshold of the computing world accounting for the majority of terrestrial communications, and in 2015, with the Expo in Milan, the spread of communication through clouds, big data, and smart objects began. Humans and objects have become symbiotic in everyday life, pulling everyone into the vortex, or rather vortices. There is the very often dynamic vortex that drags entire groups of people into the same stream, and the static vortex that is like a tornado, ready to swing open, grabbing a person up and changing their view.

The Dynamic Museums Project aims to be a model for a new culture of knowledge, especially as it pertains to the Museum in its various forms. Some cultural aspects can be very significant for DM, and therefore we must now seek to create custom users for DM, made up of the Guests and all of the people who, in some way or another, fall into the vortex of the new culture.

9.9.1 Future Positive Concept

It is evident in these times that the confident and surreal certainty of a positive and progressive future has become blurred, sometimes even essentially disappearing. The human spirit can be remarkable beyond measure in the formation of culture and the construction of those things that define a particular place and time. It can also be very strong in times of crisis. Today, we look to 2020, when the references that we have built up over the last 10 years will seem pale and often empty.

The culture today may seem like an escape. It pays tribute to a reality that is built on the past and defined by all the political, economic, environmental and social actions that every individual nation or group can build.

However, there is now a larger culture made up of all human beings, which can be visited through the network, with a huge amount of available data and information, and the management tools that ascertain the knowledge that fits moment by moment.

If you learn something, if you believe that the future will have its own reality, then your entire person becomes involved, even amidst the anxiety of an uncertain present. It reinforces individuals' desire to lean on themselves, and find themselves amplified by the reality of things, the intelligent objects around them.

It might be a good opportunity to build logic based on knowledge and participation based on Dynamic Museums.

Courtesy National Archaeological Museum of Marche

warrior

9.9.2 Knowledge, Bottomless, that Expands as It Goes Downwards

Enriched by the mass of information, criticism and cohesive data that the media give us today, we need only look back 50 years to find major differences. Our ability to know today has changed in two respects: we are more flexible and we are more willing to replace old knowledge with useful and updated knowledge. The depth of the numbers and of the data is immense, and indeed, with dynamic algorithms created on computers, you can even enlarge the diameter of the well-known.

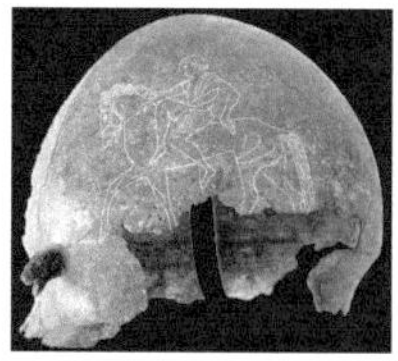

Courtesy National Archaeological Museum of Marche

Helmet

You can enter a billion pieces of information that had previously been lying in deposits. We have all watched the cars with cameras that photograph almost all the streets of the world to be reproduced online, and thus also in books, encyclopedias, papers, etc. Traditional libraries in Italian universities, understood as archives of books, have remained the primary facilitators for the pleasure of handling and reading actual volumes, but they are simply no longer the primary instruments for accessing knowledge as quickly and comfortably as possible. In conclusion, with an efficient network and a discrete computer, you can build a wide range of knowledge, which continues to widen from the surface down. DM wants to follow this growth and absorb the most attractive, and most human, signs along this journey without end. It can offer the most significant cultural experience, adapted to the logic and emotion of the individual person, and merge it all into a foundation for all forms of study.

9.9.3 The Laces of Cohesive Knowledge

Knowledge today continues to be spread, to a significant degree, through publications, books, articles, photographs, etc., pertaining to all the same past and present history that we can find on the net, and with which we prepare ourselves. Still, in that capacity, it is missing a fundamental freedom: the ability to get to learn something immediately. We are never total masters of the streams and rivers that lead us to knowledge. We consider such information to be temporary, because our memories of it lack the quality and the exact wording of the source of knowledge. We are tied to strings, or even chains, that block the source of knowledge, and it is still all too easy today to become slaves to bad information. DM intends, through its artificial intelligence drifts, to score and qualify the routes and sources of information on Museums of culture, to the point that the culture of a country can rely on these settings, since verification and certification of the knowledge can erase all fears about the erosion of security.

9.9.4 Knowledge as a Personal Gift Until 1990. Then Knowledge to All, but Selected and Attended by Few

Up until around 1950, the wealthy student could go to the library, which housed its huge store of books in an atmosphere of opulence, with valets who would seek out the required book and hand it to the young student or scholar or wise elder. It was a closed circuit, almost aristocratic, based on silence and a chosen knowledge, almost religious freedom and secular terms. A mere several decades later, you can access sources a million times richer in a short time, and explore the world from your home, on the beach, in a pub, or anywhere in the world. Anyone today can explore the world freely, and only a few works (for example, in Milan, the original codes of Leonardo da Vinci) are sacredly preserved, and rightly so.

DM offers participation in this universal experience, and transforms it into a source of pleasure, satisfaction and love for the things of life.

Courtesy National Archaeological Museum of Marche
Amazons

9.9.5 Personal Emotional Occasions Today. In the Ancient World, It Was Only the Exalted Experience, as in the Stendhal Syndrome

The Dynamic Museums Project is built to offer a choice of paths, suggested per the wishes and emotional perceptions of the Visitor, to see and hear the beautiful things among infinite possibilities. Thus, it is a time- and energy-saving device for arriving at an acceptable solution. This ability means to anticipate the visit, to plan the path, to commit exact time and resources to an acceptable goal, even to outline predictions about the experiences that you can hope to acquire. Today, these are offered only by DM. They kill the anxiety and distress, almost a syndrome unto themselves, which can bind the visitor, if the objects of the visit are still locked up in an emotional fog, and not well explored by those who know them and who have experienced them, in the Museum, as protagonists. DM will become a key for integrating the Museum and tourist routes into a natural and spontaneous cultural, personal and unique trip.

9.9.6 *The Dream, the Ambition: A Universal Museum Genome to Be Built Step by Step*

The vision of the manner in which today's culture ought to be built is very different from that which existed in the years prior to 2000. The space occupied by libraries, Museums and art galleries, representing islands of knowledge that allow one to know the culture, filled with objects and ideas, is no longer seen as the most essential, because after 2000, the immense space of digital data opened up, real, bodily, physically no less valid than the objects.

Courtesy National Archaeological Museum of Marche

Red figures

And thus was the gift of culture opened up to everyone, with everyone finally owning their own little "plot of land." Everyone could now live the reality of the culture and feel its power and value. Without the gift of knowledge and learning, physical or spiritual, theoretical or practical, it is difficult to be truly aware of one's place in the universe, to be a conscious part of one's own life.

This immense amount of largely fluid knowledge, composed of many elements that are infinitely different and profound and suggestive for humans, continues to advance and grow like a magmatic wave.

Culture today is all about listening, reading, paying attention to, and generally coming to understand the very DNA, if you will, of the people who have discovered and can express the beauty and strength inherent in a participatory life.

In today's terms, post-2000, this means knowing how to be present and how to participate, sometimes without a great amount of knowledge, but with the certainty of this very true, very *real* reality, which is the only one that exists in the present.

Courtesy National Archaeological Museum of Marche

Warrior with horse

Today, culture means the courage to accept this.

Looking at the evolution of stories, we can say that it would be different without the sense of the dimension of time. The dimension of time leads us back and forth as it wishes, and we are as children, attached to a carousel that we cannot turn. We are involved in the present. It is a bottomless pit, in which we find new parts of ourselves.

Human beings, as required by the breath of culture, immediately transform the knowledge of today that comes to us, because we do not accept, on either a moral or emotional level, the information imposed upon us by others. Human life, although now longer, has taken on a narrower dimension, and involves the individual more and more. The notion of an individual among 7 billion people may be hard to grasp, and yet, amazingly, that individual is still unique in its existence among others.

The sense of humanity is the need to breathe in culture, with its pains and ambitions, fears and moments of courage, with sudden illuminations and the joy of discovery, with the anguish of death, or the peace of order, even knowing that, with time, everything turns to ash, slips through the fingers, and disappears.

It is easy and commonplace to give an updated definition of culture, since it is nothing but individual collections of data, information and the mixing of each person based on life experience.

We may dedicate large monuments to the souls of all those who invented and developed the great concepts of history, all with great humility, as evidenced by their names, but truly, we only really know them through their communications, without emphasis. They invented the pieces of the new knowledge and offer ways to share knowledge of the culture with those who want to invent new projects, such as the Dynamic Museums Project.

Courtesy National Archaeological Museum of Marche

The Victory

In fact, the DYNAMIC MUSEUMS Project provides an example of continuity, realized through the magic of a data document, holding dates, information, represented objects, expression of feelings, in short, elements of lives lived and interpreted in the moment in which they were lived, available to look through and revive and live again with a single click. These are geological layers of ourselves, collected in an album of both real images and virtual images of holographic representations and expressions, created with a 3D printer, accompanied by music and sound effects that conjure up emotion, with photos, and, today, with files. These files can be thrown into the input space to be stored, weightless and timeless, with a coupled signal, a wave, a connection, between endless series of data, each with its own soul.

9.9.7 The Negative: As It Is in the Present Culture

The Museum of the DM Route project will be built for each person, per their culture. But let us be careful: the culture of today, if we use the year 2000 as our reference point, is no longer the culture of the years leading up to 1999, during which the knowledge stored was represented by stamps on books, people looking for and researching information in libraries, testing and correcting ideas through math. It is the culture that paves the way for notions and relies on the "consciousness" of the selection of the constructive elements. The consciousness has even taken hold of the soul, the character, and individual behavior, and the weight of the emotional part of the brain, which supports and often justifies the left rational part, has become more and more evident. This model of interpretation is simple, and has the great quality of being true. Behavior itself is the byproduct of interaction between body and brain, and even with the mortal quality of the body, a mysterious pudding of reality and dreams.

However, the bad, the downside, also lives within all these emotional stories, emerging from history as points at which human will was lost, abandoned, insulted. The DYNAMIC MUSEUMS Project also calls for the negative parts to be sought out and researched, so that through them, we can learn more about the positive and good, to see the human trajectory as it was actually lived, with courage and strength.

Courtesy National Archaeological Museum of Marche

Mars

We trust that the DYNAMIC MUSEUMS Project can also find the meaning of death in these smart objects, and discover the power of life.

9.9.8 The Reduction of Time

The journey that the DYNAMIC MUSEUMS Project has begun destroys the notion of lived time, as we have conceived of it in standard forms of education, participation and communication. Time, as it has been conceived of in recent years, with the new intellectual life enabled by the internet and the new techniques and devices used for communication, is a term contract, which kills unnecessary human news and meaningless expectations and points to a real substantiality that is often non-existent. The contraction of time emphasizes the intelligent objects, the significant moments; it blunts the harshness of pain, seeking to contain time within a happy bubble, isolated and golden. But unfortunately, it does not last forever, and the virtual reality you have built dies and fades, replaced by the vapors of a psychological tragedy, a nothing that exists and yet is nothing, if we are not willing to step inside and live with the risk. The tacit and tragic theoretical happiness of philosophers is today contracted for the purpose of inventing a mysterious time for each of us.

9.9.9 The New Dimension of Communication

It may seem shocking that a phrase or an unwelcome word could appear on the internet, and be read by millions of people, without any true indication of the source, and yet, in some way, this fact expresses the characteristics of the world wide web well and is indicative of the very references that give it meaning. Essentially, it can offend, hurt, and do very great damage, potentially without any responsibility, armed only with the power of words.

But this is the way it is, and so it has always been. In Imperial Rome, the statue of Pasquino became a kind of forum, allowing people to come under cover of darkness and leave anonymous notes, expressing thoughts, complaints, attacks, threats, even obscenities, in short, freedom of expression. Freedom of expression was guaranteed, and is even more guaranteed today in both the real world and on the virtual Internet.

This ensures that the expression transmitted is in today's forms of communication, which is neither good nor bad. Good was the communication of the Magi and the star of Bethlehem announcing the birth of Jesus, for which the pastors and citizens were waiting, knowing, as they did, of the birth of Jesus. But there was no reason why a comet should be herald of an excellent birth, if not for the pre-existing knowledge resulting in the fact.

The new dimension of communication is yet another factor demanding that the bowels of knowledge expose themselves to everything expressed so that it can be known, assessed and given judgment. However, it is judgment of the mode of communication that gives value to knowledge, the way in which it asks to enter the brain, and this lies in the soul of the individual who receives the communication. It is a closed loop of information and knowledge, and thus represents the acceptance of communication.

This is not like the herald in the square who communicated the message that had to be accepted as a decree or order or declaration of execution. Surely, someone somewhere is being overwhelmed by the insane nonsense of people as unbalanced as they are, and the risk that may result from this is present in the lives of everyday existence.

9.9.10 The Stellar Dimension of Love

The development of gratitude towards the emotional aspects of human life very clearly changes the sense of the pleasure of life and the gratification of love. They make one feel officially loved and appreciated, even if it only benefits the emotional life, rather than the passionate pursuit of economic and rational success. Many scholars of the past have conceived of laws and regulations that have enhanced the rigidity of manners set down by logic, denying the integrated and magical reality of nature, and deliberately underestimating the signs of nature itself. The concept of love as the driving force of life seems like a romantic notion from nineteenth-century literature (or perhaps even twentieth century). But no, the eternal dimension of love is born of thousands of millions of years; it assigns life, logic and emotion their true magic, creating a genuine feeling of being. The anticipated Museum in relation to DM in each case passes through the gates of love, which is free in its expansion between people, animals and even objects. All objects of memory are objects of deep love.

Consider the contribution of the DM Project to be a respectful and sacred recognition of participation in life and love, in their myriad forms, including the present life, which is made up of ourselves, and nothing else.

9.9.11 Economics and Software: Intersection of the Economic System and Software in MD

Certainly, the economic aspects of the project, from planning to development to application, following experimental tests in regard to distribution, maintenance and updates, are dominant, and mark the poles of development of the DM project. At the same time, with the same positive intention of applying the project to a global reality, extended to all countries, in all languages, representing all cultures, it has now been determined, by Spring of 2017, that this enormous and hopefully pervasive new form of culture, the biggest that has ever existed in human history, is feasible to be distributed to the entire planet, through realizable commitments that require no superhuman ability, but only technical know-how and intelligence.

Courtesy National Archaeological Museum of Marche
Warriors

However, one basic rule should be respected, established by the economic realities of cognitive development and the means of cognition: the brain of the software should fit comfortably inside the vessel of economic reality, with no danger of rattling around or even disappearing for being so small within the vastness of the box. A natural balance of economics and software, practical and intellectual skills, is the foundation from which we will soon explore a new cultural reality, to which the Dynamic Museum project has contributed.

Part IV
Social Learning Through Dynamic Museums

Chapter 10
Gone with Emotion

10.1 Small Treasures

The DYNAMIC MUSEUMS FOR SOCIAL LEARNING Project offers new benefits in regard to the social and cultural aspects of the use of Museums. Social and economic aspects are at the forefront of development of the application of the DYNAMIC MUSEUMS FOR SOCIAL LEARNING Project, because the vast and rich cultural, historical, artistic and social heritage of the museums can be revised and designed through new methods of use that will make the Museums more exploitable and allow them to excel.

When we speak of emotions with regard to the app, what we mean is the following: that we want to live the best life we can in these years marked by the fatigue of work, through efforts to create new inventions, to spread global knowledge all over the planet and to find the opportunity to participate in the totality of events. In many respects, these are the incredible gifts of reality. We can participate in this affair in an emotional way, knowing that these gifts were not present 50 years ago, or even 10 years ago,

We can carry all the pictures from a museum, 10 museums, 100 museums, in a portable object such as a cell phone, and still have enough room on our electronic sites to see each object enlarged, rotating, from different points of view, or even changing facets of its physical appearance to see how it looked originally. In fact, this visibility increases the participation and appreciation of the work of art itself, combined with the ability to include as much information about the object on the internet as we see fit.

Hence, we can also color the emotions of the object we are visiting and add our feelings from that moment, our emotions, our comments, and even—why not?—sharing them with the people who accompany us. This is a great step forward: it is a way to participate with the greatest respect for sharing emotions with other people.

It is also significant that we can even use this gift a day later, a month later, a year later, emailing it, or even printing it out and mailing it to share. We can describe our experience in that museum, adding details.

A. Rovetta and E. Rovida, *Scientific Knowledge Communication in Museums*,
Research for Development, https://doi.org/10.1007/978-3-319-68330-0_10

The visual documentation, graphics and sound become a sacred object within our knowledge, because they represent a moment in which we climbed a new step in our emotional and cultural life.

The hardest part concerns the connection between the DYNAMIC MUSEUMS FOR SOCIAL LEARNING Project and the industrial reality that manages the cultural industry. The governmental, social, economic, and institutional realities are of concern in the management of museums belonging to States and the international community. We start with a careful analysis of the various objectives of the industrial reality and governmental/institutional/social economy. The project regarding use, exploitation and acquisition of profits by the culture industry acknowledges that culture itself is only a source of economic development. The industrial reality is focused on applications that grow pervasive economies. In fact, the monument, the work of art, the lure of cultural events, these are only a poor immediate source of income and revenue, useful only for sustaining the culture itself. Intelligence, therefore, is predicated on understanding what the models, methods, and settings are that can lead to economic growth, financial development, and real pervasive social welfare.

The basic rule is that the work of art expresses memory, interest, love, and emotional and cultural participation at any age and in any social environment. This is the goal of the DYNAMIC MUSEUMS FOR SOCIAL LEARNING Project, that is, make the work of art and the event, the historical relic and creative design, more and more accessible to every person.

Courtesy National Archaeological Museum of Marche

Beauty

SIMPLICITY

If we turn to the possibility of new projects in new areas, where the new is nothing more than a step towards greater knowledge and wisdom and personal and collective happiness, then we must be willing to open new avenues of thought. The social and economic benefits of the DYNAMIC MUSEUMS FOR SOCIAL LEARNING Project are endless. The first advantage is the certainty that you will grow in terms of both conscience and culture, not with a sterile kind of knowledge, but with knowledge that allows you to be an active participant in reality, especially human reality. The social advantage is that every moment of your life will become an opportunity for cultural growth, for development towards an existence at a more enjoyable level, with more and more to share, and more and more to love. The economic advantage is without measure, where the economy is based on the social dynamic of living with others, getting the best resources out of the nature of human reality, and building a future in which your relationships with others, and with yourself, is always happier. The DYNAMIC MUSEUMS FOR SOCIAL LEARNING Project offers users, and all others who want to take advantage of what it has to offer them, indices expressing personal values. The nature of things today offers us the opportunity to inject knowledge into everyday occurrences and make up for any possible shortcomings in the information, expanding the new world through our hunger to know, to participate. The DYNAMIC MUSEUMS FOR SOCIAL LEARNING Project can be easily operated with a simple phone, allowing us to utilize our knowledge to make us feel more and more involved in the reality in which we live. Accordingly, the social and cultural benefits and rights are strong for the user of the DYNAMIC MUSEUMS FOR SOCIAL LEARNING Project, which represents a bridge between objects, between the smart things of the past, the present, and the possible future for the benefit of the soul and the wisdom of each person. The transition into this growth is entirely up to the individual person. This gift is an asset that the project operator owes to those responsible for management of this communication.

The cultural aspect is an aspect of enrichment of life. In the 2020s, the creation of personal relationships will become entwined more and more with the creation of interpersonal relationships, often only through computers, sometimes even without human knowledge. Today, we find ourselves close to the year 2020, the beginning of the decade in which the DYNAMIC MUSEUMS FOR SOCIAL LEARNING Project will take us inside our souls, collecting data and information from the clever things that surround us for our benefit. The economic benefit is another intangible but substantial advantage. The DYNAMIC MUSEUMS FOR SOCIAL LEARNING Project will become a product that serves the soul, delivering inner happiness, pleasure and enjoyment, peace of mind and emotions, all personally controlled.

The tool that the user of the DYNAMIC MUSEUMS FOR SOCIAL LEARNING Project can build for their personal use is also sold as being necessary to each user. There is the option of increasing the value that the individual can hold towards the culture of objects. The integrated through-line of knowledge of life that stretches from the past to the present to the possible future represents this value and remains a priority in the memory, in the "big data" that makes up the person's mind, that was once, in the past, represented by the library and the collection of books often read and no longer opened. But now, it becomes a dynamic treasure, to which one can gain access at any time, and from which one can extract even more value by watching and reliving emotions and feelings attached to things, to intelligent objects and events. This treasure is one to defend, its emotional world to be saved, to be enriched with Virtual Reality, or Augmented Reality, with new holographic systems and 3D printers, indeed, with any form of technology and knowledge that can involve objects, transforming even virtual reality into a substantial reality. Therefore, what we propose is to provide an innovative model for the valuation of knowledge that can take all of the media information acquired during museum visits and personalize it through the indexes provided by the person at the beginning of the visit. The objects become stronger, because their emotional values with the maximum significance indices are put to good use.

In this way, through use of the service provider, in a respectful and profound way, a treasury of knowledge is created for the Visitors that they can tailor to their own liking, and that forms a new reality. It replaces the library of books. The trove is also upgradeable, and it can be saved in big data, developed, and recovered. It is exchangeable with other people. The treasury can become a new unit that defines the person in a clear way, because every museum object is linked to its reality in time. The Visitor becomes the ruler of an empire of knowledge that inspires something in him, that recalls feelings and thoughts. The trove opens up a horizon for a world of knowledge that he/she likes, that he/she loves and that re-opens horizons of pleasure and profound feeling. That virtual treasure is real.

10.2 Overture to the World

The DYNAMIC MUSEUMS FOR SOCIAL LEARNING Project radically transforms the use of Museums by involving the personalities of Visitors through logical cognitive aspects and emotions, expressed in daily life by the respective actions of the left and right sides of the brain that form behavior.

The DYNAMIC MUSEUMS FOR SOCIAL LEARNING Project radically transforms the use of Museums by involving the personalities of Visitors through logical cognitive aspects and emotions, expressed in daily life by the respective actions of the left and right sides of the brain that form behavior. The DYNAMIC MUSEUMS FOR SOCIAL LEARNING Project boosts interest in the convergence of the Visitor's logical and emotional sides, with the participation, by computer, of the curator of the museum. One can use computers, cell phones, tablets, or any other communication device, as well as a location within the Museum itself.

Courtesy Museo Marinaro di Camogli

Faith

A WOODEN STATUE FROM ATOP A SHIP

Participation becomes a compelling thing for the Visitor. We are creating a project that unites the consciousness and emotions of the visitor into a small or large group, depending on the program, featuring the objects of the external world, in which he/she wants to participate. In addition to the world of the objects, the Visitor is also involved in the events. He/She becomes an active and participatory character in various events, visiting the entire rich, documented environment and gaining the natural opportunity to become richer in knowledge and understanding, of both the world and himself/herself, by the end of the visit. The second gift that the DYNAMIC MUSEUMS FOR SOCIAL LEARNING Project offers to the project developer is precisely this enriched awareness that is connected to all of the emotional information with which the visitor can associate. Programs are connected with external interests, to friends, to possible matches, to entertainment, to tourism, to an interest in cultural and personal emotional events outside of one's current personality.

We call this ability “opening oneself up to the world.” It lets us affirm that even a simple visit to the museum, established by the DYNAMIC MUSEUMS FOR SOCIAL LEARNING Project, can provide an opening to the world, the size of which is dependent on the size of the provider of the total project. This is an important point, to be reported as another form of enrichment offered through the DYNAMIC MUSEUMS FOR SOCIAL LEARNING Project.

To put it in a single expression, we can say: OPEN YOURSELF UP TO THE WORLD.

10.3 Castle

The DYNAMIC MUSEUMS FOR SOCIAL LEARNING Project uses very basic mathematical algorithms. However, it handles them in a distributed way, well connected to the Visitor’s character.

1Courtesy Collection AR

Open Castle

SCIENCE AND TECHNOLOGY TODAY

By means of mathematical algorithms, the DYNAMIC MUSEUMS FOR SOCIAL LEARNING Project generates a numerical-physical platform, suggesting results to the Visitor, listing the most pleasing and interesting museum objects in each of the collected museums that suit the visitor’s intent. Additionally, we can add one other very important aspect: the customization of the project. The algorithms that the DYNAMIC MUSEUMS FOR SOCIAL LEARNING Project uses are very basic and simple, and it handles them in a distributed way, well-connected to the Visitor's character, that enriches self-knowledge. This project is dedicated to all indiscriminately, so that everyone may gather around the object that is of most interest. No, let me rephrase that: this project is dedicated to you, the Visitor, and makes you the center of the universe. You collect this universe and decide what you like best, to grow and complement your human nature, your wisdom, your love for people and things. The key point is that, for you, the DYNAMIC MUSEUMS FOR SOCIAL LEARNING Project builds a wealth of information that you can access. It is organized mainly to replace those libraries of books, so that people need to visit libraries less and less, because you can get into the book of life, of history, of knowledge, of your knowledge in the world. The DYNAMIC MUSEUMS FOR SOCIAL LEARNING Project, with the consent of the project coordinator, will open up a wealth of information in which you are the protagonist. You are a fast walker, conscious and always self-aware. The project prepares worlds of images, sounds, and information for you, which are easy, simple, friendly, with you in the middle of it all, if not, perhaps, your face and your body, and you are the only person capable of understanding and deeply appreciating what is provided. Let us set it out as such: the gift is a set of objects that we have chosen, but that you have built, and they are steeped in meaning that relates to your nature, your past life, your very being. Nothing has ever given you such a strong adherence to your present life, and a love so open to your future. You are fully aware of the "house" that provides you with this help, that allows you to open your database, your photos, your letters and your entire lived history.

We call this gift received wealth, because it is easy to remember the most beloved moments. The Visitor must approve individual packets, re-evaluating and possibly changing them, and, if you like, deleting everything you do not believe should be used, because, as the Guest, this property is under your complete control.

We have added three riches to the DYNAMIC MUSEUMS FOR SOCIAL LEARNING Project. One is the treasure; the second is the sharing of the story; the third is an internal element, very intimate and important.

We now want to add a fourth gift. Its generation is linked to an emotional baseline that characterizes the visitor in his/her drive towards an emotional reaction to the Museum visit. This fourth fundamental point concerns the possibility of creating a new castle of understanding, alongside the canonical system of the DYNAMIC MUSEUMS FOR SOCIAL LEARNING Project. In fact, the program includes 16 logical and emotional elements indicated by the Curator, with one among them holding the highest rating. The program can take this object with the highest score and apply it to all elements of the adopted system of Museums. The result is a selected set of objects, one for each museum, that are placed into the basket holding the Visitor's objects, based on the fact that they respond to his/her favorite aspect.

Courtesy Collection AR

International Teams

UTILITY OF INTERNATIONAL TEAMS, IN EVERY ASPECT, FOR THE BEST COOPERATION IN DESIGN, ACTIONS, DEVELOMENTS OF SOFTWARE AND ARTIFICIAL INTELLIGENCE TOOLS

As an example, suppose that users choose the most important aspect for them as being beauty, with a score of 9 points out of 10. The program can apply this 9-point beauty index to all elements of the museum circuit, one by one, and each user, per the instructions of the software editor, can then choose their favorite items from each museum. If we have 725 museums, for example, we will have 725 favorite objects; if we choose, out of those 725, only those characterized by a specific interest of the visitor, the software will select fewer items. Each preferred object within this "Castle" is linked by a ranking of the assessments of Trustees of each museum, different, infinitely different, from each other. All these objects that match the index preferred by the Visitor are important, because they are the result of the collaboration of all the curators, which gives rise to that object being chosen by the most Visitors as being pleasant and pleasing. The "Castle" will therefore have a ranking, with many objects of different scores; intelligent software will help the visitor to choose the ones that they like best. In conclusion, the "Castle" of each visitor becomes full of valuables, with ranking being dependent on the users' favorite principal value.

All of the objects gathered in the "Castle," in which the visitor is the protagonist and the owner, correspond to the Visitor's main desire. The creator of the project must choose the best possible curators or define new tour guides who can assign logical and emotional values to selected objects from the museum project through creation of an exceptional and uniquely efficient circuit. In this way, Visitors, using the tools of the DYNAMIC MUSEUMS FOR SOCIAL LEARNING Project, will build a "castle" of their property, becoming part of a network that enjoys the use of new scientific and technical realities, such as three-dimensional printers, objects in virtual and augmented reality, with neuromotor and sensory feedback, which will become even easier to obtain the closer we get to 2020.

10.4 New Habit

Now, it is time to cross the new museum dynamic and progressive culture with all the new knowledge and new characteristics of life. In this millennium, a new cultural world is being driven, a new life for every person, touching upon the inner life of every individual, whatever their cultural level, through the most advanced technologies. Today, we get the most amazing results, The most intense passions towards knowledge of the past, the present and toward the leap into the future are being born. We can order machines to move, the pens to write, and we humans to interact with all the tools in our daily life. Even one's tone of speech and expression can affect your computer, which is now able to recognize them. We also have the chance to feel the nuances of voice and understand the state of mind of the person. New horizons for understanding, participation, the shared emotion of people interacting with each other are appearing.

The power of modern means of artificial intelligence creates the possibility of saving our museum, building a story in an almost automatic way, a path with figures, sounds, and words that we can review as related to our museum visit, and remember the emotion that it stirred in us.

So, the Visitor becomes both a builder of his own mind and a navigator of the past, so that cultural growth, which, as is well known, really takes place in steps, continues internally.

Who does not remember how it feels to find oneself standing in front of an exciting work, an object of such significance that it arouses memories, impressions and dreams? Is this not an unforgettable experience that becomes etched into the cortex of the mind, in the conscience and in the heart?

It is true that technologies used to realize communication systems often do not prevent noise and malfunctions.

However, the wealth of documentation, figures, drawings, and photographs of the works becomes a memory point that can be saved in Big Data and communicated, transmitted, and shared socially in its entirety.

The Visitor is not only the protagonist of his visit, but also builds and realizes it as the object of his experience and his life. This element, which corresponds to a file, is saved, along with the date and any photographs deemed necessary (and may also include a selfie to certify the day). This is a unique concept.

10.5 No Longer Forbidden

In 2010, in Beijing, the Forbidden City was opened to visitors. For the first time, anyone could enter into all of the interiors, through use of a large electronic coupon in the form of a card, featuring GPS, LEDs and a micro map that identified where the visitor was at any given moment. It was very convenient, because it made you feel immersed in the halls, courtyards, squares, and gave you a sense of belonging. Along with the GPS and LEDs, the electronics within the card were also used for the storage of information. In many museums around the world, we have applied computer systems with all communications applications attached to cell phones, touch screens and all other devices that allow you to communicate. In five years' time, Museums around the world have felt the shock of the spread of computer use within their context, although these devices have not permeated the entire context of the museum.

Courtesy Collection AR

Technologies and History

VISUALIZATION AND GPS

The display of works in a Museum has been tied into the Internet ever since there was an Internet; in the early days, a visitor might use mobile phone, but most likely in secret. But slowly, over time, use of phones has become pervasive in museums, with photographs and home movies of the visit becoming standard for Visitors,

Which are the most important applications of this type in regard to museums?

It can be argued that this started with the Metropolitan Museum in New York, which was one of the first to allow the use of cell phones to take pictures for documentation in the museum.

This new strategy is important, because the visitor becomes capable of storing, saving, and thus remembering the love he/she felt during the visit to the museum, ensuring a well-informed and significant museum trip. Other museums in the world have preferred to deploy a mobile app acting as a guide for the Visitor.

The visitor therefore has aural and visual support that supplements the educational and cultural content, for complete help in recalling the experience. This is a great start, but the DYNAMIC MUSEUMS FOR SOCIAL LEARNING Project seeks to take it further, building the entire project on a platform that is tied directly to the scientific direction of the project, and thus can be modified, adapted, and all the extra changes made available within the Project.

Apps, generally speaking, include images, some explanations, a series of photographs and a link. They can really be of help to visitors, but in practice, they are simply a different configuration of the guide at the Museum.

We can supply a basic list of major museums that use apps as their primary electronic guides.

The basic difference within the DYNAMIC MUSEUMS FOR SOCIAL LEARNING Project is in its completely different approach, since each object is evaluated according to both the interest, curiosity and love for knowledge of the visitor and the culture and wisdom of the Museum Curators.

Clearly, an intelligent future lays in store for the museum tour, led by the DYNAMIC MUSEUMS FOR SOCIAL LEARNING Project, one that can address individual museums in a sequence of museums in the same region. The global discourse on museums proposed for visitation will take place.

For each object, through today's electronic communication, it is possible to attach a picture, a story, a cultural event that is also linked to the Visitor's character. It will be palatable to the individual's knowledge, their culture, and their system of appreciation of reality, all of which will become further solidified through the entirety of their museum visits.

Hence, Museum visits will become a great treasure, a gift to human nature, to intelligence regarding the reality of human history, as the visitor moves to the center of the Museum visit and becomes arbiter of his/her own new knowledge.

This is the richness of the DYNAMIC MUSEUMS FOR SOCIAL LEARNING Project.

10.6 Advertising and Publicity in Dynamic Museums

10.6.1 Feeling and Communication

The advertising for a dynamic museum is linked to the feeling that the museum, with its objects, its visions, its projections, its touchscreens, its simulations, must engage the visitor. The museum's "pull," to use a term from advertising, must flow from its objects. These are, as we know, the essential elements. The museum object must be of interest for its beauty, must call to the person, must fascinate the visitor. However, the decisive factor that makes a museum memorable is the necessary intellectual entertainment, a reference to the facts of life, an incantation from the object that tells its story.

In advertising the object, an optimistic desire must be expressed to be known and loved by anyone, even those of other nations and cultures.

The object, to be accepted as a point of reference, must be simple to understand, simple almost to the point of being representative of a smile that invites the visitor to take a closer look. The appeal to people of different cultures and spirits must be born of an attractive force. It must be considered that the statue, the object, the book, the work of art, have their own individual history, and visitors should be able to feel confident that they will be helped in learning about the events behind the object.

We can describe the advertising of the Dynamic Museum in two ways.

The external advertising, which takes place outside of the museum, is entrusted to the internet, to the communication networks of the various regions, linking the objects to the museums in which they live through computer networking applications. The second approach to advertising that the museum can take is to make the visitors feel as if they are involved in the cultural discourse in an atmosphere of cheerfulness, simplicity, and lightness, making the visit an enjoyable one. The museum is not a catacomb, the museum is not a church, the museum is not a place of induced holiness. The dynamic museum today should be a pleasant place, in which one would not even be surprised to find a restaurant, a bookshop, or a store. Visitors can easily enjoy moments of rest, before, after or even during the visit. This form of cheerful sharing of the museum is the best possible approach, because it will turn the visitors into a beauty ambassador who tells their friends and relatives about what they have seen and experienced.

But even better, these sights and experiences shall then be disclosed to the whole world, because every single grain donated to knowledge is a gift to the entire culture, enhanced through emotional participation and social learning.

Advertising can approach the object from two different perspectives, one outside, one inside. The first is as a participant in a story that helps the visitor see the object as a living thing that had its own intelligent experience. One example of this is Garibaldi's octant at the Maritime Museum of Camogli, Garibaldi kept it with him throughout his life until his death in Caprera. The second way is to focus on the object as a thing of beauty, almost mysterious, which lingers in the emotional memory of the visitor. It participates in the social discourse of knowledge and learning that turns culture into a pleasant game.

10.7 Religion and Museums

Since 2017, museums have started to become more dynamic in their social and emotional aspects. In regard to religion, appearance is most critical in today's intellectual development, because it slips between politics, science, and culture, and must always respond to the most profound needs of the individual visitor.

Courtesy National Archaeological Museum of Ferrara

Vision over the Earth

Faith religion love self-consciousness

In some places, religious communication and fideistic participation are increasing in daily life, though sometimes concentrated in religious events.

The museum thus becomes a collection point of past events. Objects of faith or those otherwise referential also offer the possibility of revisiting episodes that touched the culture and affected human sensitivity. Museums generally provide the ability to manage religious events in a social way, since art and history often represent the bulk of human heritage, which often involves faith. Communication as a source of faith is getting stronger, more intense, more evocative. The message of faith is invested in by millions of people. The museum is only a cultural meeting point of civilization, of faith, of thought with the soul. The museum is a collection point of the history and future possibilities of every religion, because where there is beauty, the simplicity of the culture is clear and is expressed through application to the individual. Museums and temples can also share sacred objects, witnesses of religious development.

10.8 Fashion, Creativity and Dynamic Museums

10.8.1 Fashion as an Indicator of Daily Life

Fashion can be an indicator of everyday life, even in museums, most often as a witness to the times. There are also splendid Dynamic museums dedicated to fashion, such as the Prada Museum in Milan, where the work is associated with its cultural impact and historical beauty. The Armani Museum documents the story of the Armani suits, in all their style and beauty. It retraces the development of the suit step by step throughout the museum. Fashion itself is a facet of all of the objects featured in any museum: from ancient Papuan costumes to Roman dresses, from Hawaiian clothes to Greek attire of antiquity, and so on. All of these items, and many others, are documented by the museums not as a documentation of fashion per se, but rather because they represent documentation of everyday life. Museums have a cathartic effect, because they are the indisputable documentation of how the human dressed, including various forms of jewelry and accessory. Museums have been known to display antique jewelry that modern jewelers then refer to, by copying and redesigning them in a modern way. We face a time when the museum can basically serve as a giant document, beautiful and powerful, giving account to the creativity that evolved over thousands of years, from the ancient tunic of Roman engravings to attire featured in paintings from the Renaissance to modern clothes. Today, fashion has become accessible to billions of people. Fashions become known around the world through the internet, and anyone can be involved in one way or another, although not always as a purchaser, given the high cost. While museums that seek to contrast the Ancient and the Modern serve as documentation of creativity, in regard to fashion, the museum becomes a mirror of reality, as the fashion itself is filtered through the user of the object. Today, fashion is a quality of life, and therefore we must consider this creativity in dress as an element that can bring you closer to history through the museum's representation. The biggest fashion creative experience took note of the first clothing, fashion items, dressing, shoes. Museums allow us to develop intense sensitivity towards what the customer is feeling. The dynamic museum becomes an incredible exercise in witnessing the development of creativity. Today, a transition from fashion to design, and from design to construction, is something that is very frequently seen. The transition is very natural to the creative nature, and is stimulated by new courses of new schools that want fashion to be a substantial aspect of the enrichment of human behavior.

10.9 The Emotion in Museum Visits

10.9.1 Sharing

Sharing the modern experience of visiting museums that are networked through the use of sharing tools allows visitors to both view and participate, sometimes through augmented reality, virtual reality or other forms of sharing. Today, we not only have books and catalogs but also, through the 3D printer, reproductions of beloved objects, made to scale, that the visitors can take with them when they leave. Love and sex are behaviors that define a person's life, just as much as their mode of logic. The museum cannot fail to take this into account, and should thus also be prepared to interpret these desires and wishes in relation to the human manner of existence. We must have museums that are designed to take into account the realities of modern, everyday life. First, because Visitors visit museums all over the world, and second, because the visitor's acceptance must be total, with the museum responding to their needs.

Courtesy National Archaeological Museum of Ferrara

Emotions

DESIGN , CREATIVITY FOR EMOTIONS

The preparation and presentation of exhibits cannot take into account every single emotion from every single religion and cultural setting. Today, we truly are faced with revolutions of immediacy, both cultural and social, involving billions of people in the blink of an eye. The phenomena of positive communication and negative terrorism can invade the planet in seconds. This was not the case thirty years ago, twenty years ago, not even 10 years ago. We must consider that during a museum visit, a couple is faced with presentations of intelligent "Objects" to which each individual responds differently, and the feeling is generally surprising, because culture always affects the entire personality. Not infrequently, a couple may have different individual reactions. It opens a new challenge for the museum, to present a culture to the female part and to the male part. Accordingly, integrated parallel paths are offered to the different groups that show different approaches to the objects.

10.10 Treasure of Space Exploration for Museum Participation

10.10.1 Concrete Challenge

Today, space exploration is the new great challenge, with the landing of the rover on Mars, the construction of villages in a space station on the Moon, construction of waiting stations for later human exploration of Mars, not to mention all the new automation related to space travel: rockets, orbital sightseeing, zero gravity launches, the capture of asteroids in space for the purpose of altering their orbits, the exploration of said asteroids and the exploration of the Universe in general.

Recall that today's human only truly knows 5% of the space in which we live. The other 95% is a mystery, pure mystery.

The approach of museums to space travel can be divided into two parts. The first constitutes written documentation, composed of important, fundamental texts, some of which even date back to the Renaissance and indicate some thoughts and certain realities regarding space that had already been intuited at the time. The other approach involves museums that house components of vehicles that have flown in space, representing the aspect of science and culture development.

The most important museums for space research are in Los Angeles, Pasadena, Toulouse, and Beijing, as well as others in Germany, Holland, Belgium, and Italy.

It seems important to indicate now that space research is uniting all countries and nations, from the United States to Russia to China to India to many nations throughout Europe; Countries in South America are making contributions.

The museum must create an interactive form of speech and communication, using anthology sources. It is therefore proposed here that a network of museums be created of manageable space content, enhanced with virtual reality, which would also offer simulated physical exhibitions, with remote-controlled simulation. These

Courtesy National Archaeological Museum of Ferrara

Treasures on Earth

CONCEPTS OF INFINITY INSIDE A ROUND PLATE

would serve not just as history museums, but also as a propulsion system for knowledge about space.

Today, we have the ability to connect all the space museums and institutions, and it could very soon lead to the establishment of a new museum on space that truly emphasizes the prospects for the future.

It is very important to have cooperation in development of the proposal, so as to be able to present the idea of connecting the main space agencies naturally.

There are museums that house very important objects of exciting beauty, some of which are located on bases involved in space travel, in collections contained in their pavilions. We accept that space can offer its beauty to museums, which would then become interpreters, owners and capable ambassadors.

Is this the future in which we could be the possible protagonists?

10.11 Volunteers, the New Economy and Museums

10.11.1 Economic Structures

The third consideration is that volunteering is encapsulated, organized, and structured within the global economy. Since it began many years ago, we have followed the intelligent voluntary birth of active and functional economic structures, particularly in countries like India. Now, in 2020, we are going to find ourselves in a new situation in which these economic structures, which have grown enough to reach even the big banks and financial institutions, can be sources of cohesion between the business world and operational creativity. What contribution can a museum make in this respect? A dynamic museum can make a great contribution, because it can be a witness of what is happening and explain why banks and institutions all preserve the history of their activities.

Courtesy National Archaeological Museum of Ferrara

Creation of amphoras

HISTORY OF ACTIVE LIFE IN MULTICREATIVE ACTIONS

For example, in Milan, there is the Martinitt-Stelline Museum, dedicated to orphaned children, maintained by public funds. It is a very nice and contemporary museum that uses the latest technologies for communication, including virtual reality and augmented reality, touchscreens and many useful means of multimedia communication. The museum documents the effort, the problems, and the difficulties inherent in creating a reality that allows kids in a difficult situation to come together and blend with each other.

This third consideration can become a challenge for the museum, serving as an intermediary school linked to reality. In a dynamic museum, **Curator must learn what changed; it is not simply the Visitor who must be updated and cultured**. People need their knowledge to be updated, and this is the real goal and task of the dynamic museum. The third consideration means collaboration in profit in a much wider sense, not simply economic and localized.

The reality of a new design for museums is a challenge that the DYNAMIC MUSEUMS project will have to face.

10.12 Soul Sign

Talking about a museum's objects means talking about things with soul. The most common term in use today is "smart," an idea that indicates that objects have a story, a reality, something that surrounds them and makes them visible in regard to many aspects of intelligence, sensitivity, connection and participation. We can say that objects, despite being inanimate, become intelligent, participating, in their way, in human intelligence.

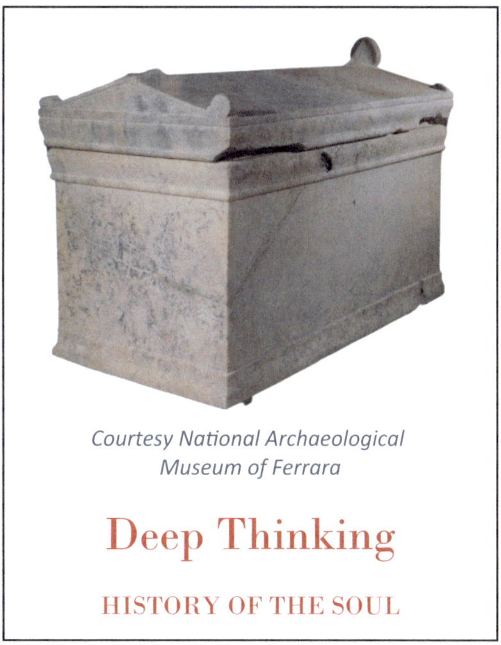

These smart objects, infinite in the modern world, can be thought of as talking objects, that is, objects that have the ability to express themselves, as demonstrated by the fact that each photograph of the object, of the person, of the environment, of the landscape, of the work of art can be associated with sounds and words that have specific meaning for the object. The object has its own sense of intelligence in the history built into it, but also its own manner of expression. If we look at the infinite environment that surrounds us and the objects within it, each of them can be associated with a memory, a word, a thought. We can say that the objects can speak, and the words are something linked to the person who interacts with them.

We must give each object within the museum a word, or several words, or a phrase, something that is born from a human being. Associated with this object, the word cuts the world in half. Today, we can keep billions of objects with trillions of words, and create more paths of thought, of love, of life, of peace, of joy, that guide us through a museum, a history, a collection of works of art. The Dynamic Museum project wants to offer intelligence and emotional management, as well as knowledge on the part of the curator of the museum. Therefore, an object may express itself and connect with us. We are moving towards complete knowledge of and participation in an emotional and beautiful life.

Part V
Features of Dynamic Museums for Social Learning in Action

Chapter 11
Project Contents

The DYNAMIC MUSEUMS FOR SOCIAL LEARNING Project offers applied development of computer systems for the enhancement of Museums. The project extends the results of the valuation of the Museums so that tourists may participate in the wealth of the territory.

The DYNAMIC MUSEUMS FOR SOCIAL LEARNING Project has already developed a high-quality platform through which each Visitor can enter and participate in its proposed program, enroll in related events, schedule trips and tourist routes fully aligned with their intentions and desires, and engage in spiritual enhancement of the territorial culture for their own enjoyment. The intervention of the DYNAMIC MUSEUMS FOR SOCIAL LEARNING Project, with the support of important sponsors and partners, deals with museums, tourist itineraries, hotel and catering conditions, available means of transport with times and durations provided, and information on all the local sports, theaters, and movies that may be needed for perfect participation in the reality of the local culture. The project therefore facilitates a more intense friendship between the Visitor and the Land, and attracts new visitors from other regions and countries to take advantage of the wealth of the territory. The DYNAMIC MUSEUMS FOR SOCIAL LEARNING Project will continue to expand, to meet the new demands of the visitors. The proposals can share the valuation method proposed by the project on a global scale.

The Project's Sponsor could be any or all of the local economic and financial entities with a stake in the cultural knowledge of the museum's treasures. Knowledge management affects all people who are touched by the beauty of the works and objects found in museums. Sponsors can provide financial support for the launch of the project, which, using applications and attendance systems, can sustain itself after startup. The development is enabled by Italian and foreign participation in the growth of Museums. The implementation of the project consists in an application that can be downloaded onto any phone and that allows the visitor to prefigure and build a potential personal program of visits to museums, with models that respond to individual interests and personal tastes.

A. Rovetta and E. Rovida, *Scientific Knowledge Communication in Museums*,
Research for Development, https://doi.org/10.1007/978-3-319-68330-0_11

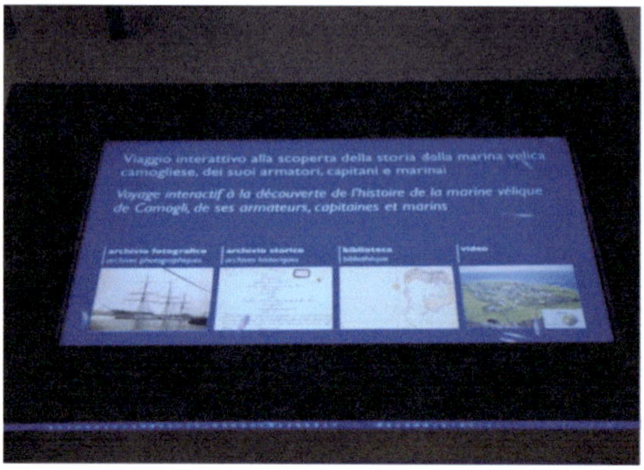

Courtesy Collection AR

Communication happiness

BLUE TOUCH SCREEN AND A DEEP DATA BASE WITH CLEAN IMAGES, FOR OUR FINGERS AND EYES

11.1 The Project's Strong Points

- The DYNAMIC MUSEUMS FOR SOCIAL LEARNING Project opens new ideas for producing new value, both cultural and social, of territorial museums.
- The DYNAMIC MUSEUMS FOR SOCIAL LEARNING Project devises and revises new methods for using the rich cultural, historical, artistic and social value of territorial museums.
- The DYNAMIC MUSEUMS FOR SOCIAL LEARNING Project radically transforms the use of museums by putting the visitors at the center, encouraging them to engage with the museum through their own personalities and the logical and emotional cognition they express in their daily lives.
- The DYNAMIC MUSEUMS FOR SOCIAL LEARNING Project boosts the convergence of intellectual and emotional interests through joint participation of the visitor and the curator/museum assistants, by way of electronic devices (PCs, smartphones, tablets, or any other communication device, including a station in the museum itself).
- The DYNAMIC MUSEUMS FOR SOCIAL LEARNING Project generates a numerical/physical platform of the results suggested by what the visitor lists, with a ranking of the museum objects that are most pleasing and interesting.

- The platform is also used to reconfigure museum routes, thanks to the feedback of the visitors, offering them a new way to enjoy through identification of ad hoc paths.
- The result for each visitor can be identified through degrees of satisfaction, which will hopefully be high enough to make them want to share the experience with other people.
- The result for Curators and their assistants is to extract themselves from their daily work routines to focus on the creation of new initiatives. Starting from the consideration of different placement of the various works, the option arises for the use of works currently lying dormant (sleeping beauties, if you will) in warehouses. The curators/assistants, their motivation increased, will also be encouraged to expand services for even more appealing fruition of the museum's offerings, using prints, photos, 3D printing, and so on, all for the purpose of providing the best possible support to the visitors and turning them into "ambassadors" of the museum.

Courtesy Museo Marinaro di Camogli

Instrument and symbol

OTTANTE INSTRUMENT FOR NAVIGATING AND HISTORY OF A NATION

11.1.1 *Objectives of the Intervention*

The objectives of the "DYNAMIC MUSEUMS FOR SOCIAL LEARNING" Project are:

1. Provision of a "DYNAMIC MUSEUMS FOR SOCIAL LEARNING" Program for the Museums' exploitation, to be used as a subsidiary of Museum activities and for dissemination of Museum objects in close relation with the cultural and emotional demands of potential visitors, including foreign visitors, for the widest possible enjoyment of the Museums
2. Heightening of attention to small museums, often not incorporated in full in Museum tourist routes, to heighten their visibility with information and communication, presenting special areas of cultural interest and Museum knowledge
3. Possibility of collaboration with institutions and foreign companies to promote the use of the DYNAMIC MUSEUMS FOR SOCIAL LEARNING Program for the benefit of cultural tourism.

In short, the DYNAMIC MUSEUMS FOR SOCIAL LEARNING program consists of an application downloaded onto a smartphone or electronic system, the program of which allows visitors to prefigure and build personal guided tours, dedicated not only to the museums, but also to other forms of entertainment (cinemas, theaters, restaurants, sporting events, and so on) in line with their interests and personal tastes.

- It makes possible the alignment of museum activities and related works in close relation to the cultural and emotional demands of potential visitors (including foreign ones) for use in measuring the works featured in the museum.
- It promotes the development of attention to small museums (often not included in museum and tourist routes) to broaden knowledge and utilization of the entire territory.
- It creates the possibility of an increase in the number of visitors, brought about thanks to the "positive word" of the same Visitors who, satisfied with their own experiences, become an effective communication vehicle.
- It facilitates the conversion of the non-frequent visitor into the frequent visitor, through the charm exercised by the content of the project enhanced by the application.
- It facilitates the increased frequency of visits thanks to the added value created by the curators/assistants through analysis of the feedback collected.
- It facilitates collaboration with foreign institutions and organizations to promote the use of the DYNAMIC MUSEUMS FOR SOCIAL LEARNING program in visits to museums.

11.1.2 Italian References

The initial experiments were conducted in collaboration with the Ministry of Heritage, Cultural Activities and Tourism in five national archaeological museums, in a catchment area extended to the whole Italian territory. The project proposes, and illustrates in practice, a cultural tourist route that also ties in local events according to the individual's emotional involvement. It is a unique approach, predominantly because it springs entirely from the emotional behavior of the visitor, who becomes the sole protagonist of the visit. The aim is also to offer a new vision of the museums, in a comprehensive assessment. The Curator of the museum is part of the critical sensitivity of the visitor, who can experience pleasure in finding, in digital form, among infinite possible exposures, the one that best aligns with his/her sensitivity and his/her desire for culture and participation. The objectives therefore are not that the program acts as a substitute for the museum culture, tourist information and events, but rather as an integrator into a new reality for the visitor who has declared his willingness and desire to participate in the culture. This multiplicity of essentially cultural goals and personal growth is a main feature of the program. The action of a Sponsor/Partner in this project can be applied throughout the methodology of knowledge and participation, for maximum enhancement of all cultural, tourist, communication and participation aspects.

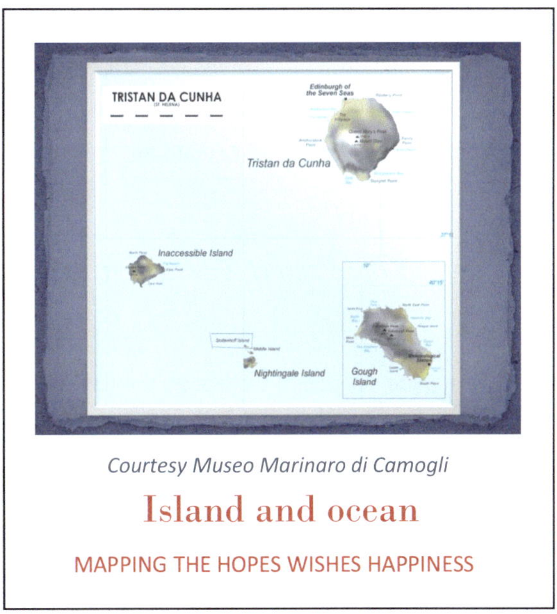

11.2 People, Times

The key players affected by the planned change will be those people who have recently attended the Museums, only to find incomplete knowledge or a lack of opportunities. Also those people who are passionate about Museums, and who are open to communication that will choose and report Museum objects to them, some of which are unknown, but all of which fall within their sphere of interest, curiosity and pleasure. Visitors and tourists will become promoters of participation in culture and tourism.

11.2.1 Subjects that Will Be Affected by the Expected Change

When we refer to passionate people in terms of the DYNAMIC MUSEUMS FOR SOCIAL LEARNING Project, we mean people interested in communication who will choose to come see those Museum objects that, while sometimes known and sometimes unknown, all fall within their sphere of interest, curiosity and pleasure, and then report on them to people they know. The museums themselves will then find new people who will be attracted by the charm of the exhibits, and they too will be partakers, offering texts, visiting tools, images, 3D prints in three sizes, etc. This will also act as an aid for promoting cultural tourism that is more suitable and acceptable to the individual person. This will also prove to be an advantage for the territorial culture, for both residents and tourists, and for those who want to see the universe of the territorial culture expanded beyond the character of its local reality.

11.2.2 Application of the Project Towards a Rebound of Italy

The first months of the implementation of the first part of the DYNAMIC MUSEUMS FOR SOCIAL LEARNING Project in the five National Archaeological Museums chosen by MiBACT has already caused a huge change in the visitors and the Curators of the museums.

We can therefore now give a realistic estimate of this change based on the data from more than 1500 opinions and Visitor trials collected in the five national museums.

Suffice to say, significant change is expected; the new confidential and aggressive system shared resulted in an average score of about 82–100. Visitors were faced with a new system, presented in a confidential and positive way by the project. This cultural reality is, by itself, already indicative of strong substantial change.

Other stakeholders will be the regions and institutions that have already offered to collaborate in building the tourism network suggested by the Visitors to the Museums through their available "connectors".

Subjects affected by the change will also be responsible for enhancing finalization of the visitor's museum experience, with special emphasis on that which will make the visitor a happy friend of the museums.

The Senate Committees for Cultural Heritage and Industry, Trade and Tourism are all very interested in this project. Therefore, we will present the results in Rome to said Commissions, in collaboration with the Ministry of Culture, for the purpose of considering development of the project beyond the five national museums.

We believe that these bureaus will also be interested in the Italian cultural reality that will result from adoption of the DYNAMIC MUSEUMS FOR SOCIAL LEARNING draft for use as an extension of its "Easy visit" project, which aims to make tourism to the area a lot more dynamic and much simpler.

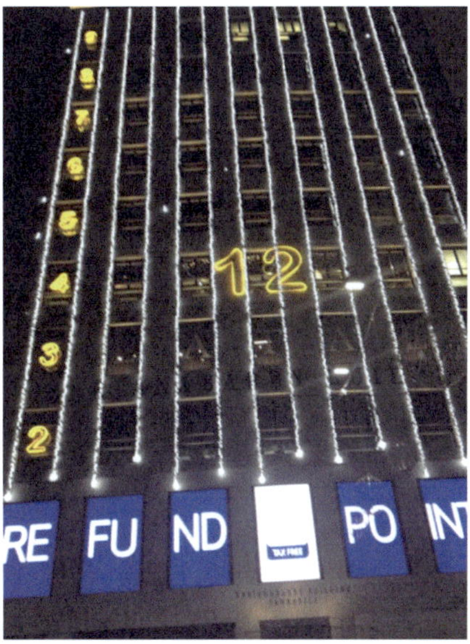

Courtesy Collection AR

Images and Buildings

IMAGE SPREAD TO THE REALITY OF DAILY LIFE FROM BUILDINGS

11.2.3 Times in Which the Change Is Expected

The change induced by the DYNAMIC MUSEUMS FOR SOCIAL LEARNING Project will be felt immediately, since the museum information will be transmitted in multiple modes through the sharing of Visitors and the collaboration of the Curators of the Museums, facilitating the preliminary presentation of the Museum objects, defined as "intelligent objects". At the same time, the overall change will be gradual and very smooth, since it will spring from the visitor's involvement in his own logic, knowledge and emotional reality.

Progressively, more Museums will enter into the virtuous circle of museums featured in the database of the project, provided by the Museums themselves. The selected objects will all be linked to logical and emotional descriptions.

None of this need be linked to a specific culture, refined and pervasive, but only the desire to feel pleasure and to satisfy curiosity by entering the museum institutions.

The change will occur at time zero. This ongoing experiment will continue running in all 5 Italian national archaeological museums, with support from the Ministry. The issue of re-proposed sharing with pleasing results and the fulfilling, active and emotional involvement of museum curators will allow for consideration of immediate changes to the Museums and lead to participation in events around the museums themselves. Do not forget that, within the global landscape, the Italian is essentially one of the most learned persons on average. It follows that increasing participation year by year in the reality of the museum is a symbol of this. The DYNAMIC MUSEUMS FOR SOCIAL LEARNING Project provides the opportunity for cultural and emotional expansion through use of the most advanced technologies. The project also provides for the availability of items made with 3D printers that can be built for the visitor immediately at the end of the visit. The project also includes the possibility to offer a hologram of the Museum object, when possible. These two gadgets make the project very nice in application results.

The lure of these gifts at the end of the museum visit, the availability of tourist information, and the information on adjacent events will make the experience more participatory and cause more immediate development. The DYNAMIC MUSEUMS FOR SOCIAL LEARNING Project thus supports the economic reality.

11.3 Business Plan

The business plan has been drawn up based on the experience already conducted in a positive manner in the national archeological museums, which has been applied to the DYNAMIC MUSEUMS FOR SOCIAL LEARNING Project in preliminary form. Economic management will be settled between the museums and institutions involved in the project. The business is linear in its development and will be updated periodically, according, in part, to the involvement of the tourism, sports, cultural and economic institutions with which contacts are established. The need for staff will be limited, because the interest that will be sparked in the museums and related institutions of the state will inspire them to select the necessary resources. It seems clear that the ambition of the project to become an asset in a new approach to participation in culture according to developments in modern neurobiology is already apparent in the structurally modified design method. Today, participation means sharing. The DYNAMIC MUSEUMS FOR SOCIAL LEARNING Project proposes the following plan for the first year.

BUSINESS PLAN (Year 1)

1. Research and Development
   (a) Development of a working prototype of the application
   (b) Involvement of more Museums
   (c) Development of specific protocols and drafting instructions
   (d) Development of programs based on previous operations
   (e) Demos and presentations generated in the field.
2. Creation of the Product
   (a) New appealing design
   (b) Using symbolic icons to make each presentation appealing
   (c) Application with 3D printer outputs, holography, etc.
   (d) Final evaluation and proposal of the application and operating costs.
3. Development of the general organization and recruitment plan
   (a) Development of computer and museum assistance in the application
   (b) Promotion and application in all major languages, increasing international visibility.
4. Marketing, implementation, promotion
   (a) Comparison on the national and international markets
   (b) Evaluation of the implementation of the DYNAMIC MUSEUMS FOR SOCIAL LEARNING product on a wider scale
   (c) Initial use of the application in Museums, then branching out to exhibitions, theaters, shows, sporting events, hotels, transportation, utilities
   (d) Promotion within the context of youth and children's entertainment
   (e) Promotion in the world at both a basic educational and higher intellectual level, through the creation of circles and participatory structures working at a high emotional level
   (f) Detailed promotion abroad of the museums and the lives of the regions within which they exist.
5. Fundraising. Creation and development, based on prototypes, of an international model for use within museums and other outlets of cultural entertainment

6. Sales and marketing for widest circulation of the DYNAMIC MUSEUMS FOR SOCIAL LEARNING Program.

Courtesy National Archaeological Museum of Marche

Archeological Treasures

TREASURES OF BEAUTY AND DESIGN

11.4 Critical Aspects of the Platform

The platform can also be used for reconfiguration of the museums themselves and for the reconfiguration of museum exhibitions, which can be organized according to the various categories of visitor and prepared for each individual visitor.

The platform's main function is to reconfigure Museums and Museum routes so that they offer every single visitor a uniquely-tailored experience.

The steps are:

1. Application

   - Apply the DYNAMIC MUSEUMS FOR SOCIAL LEARNING Project to a museum so as to create a personal journey for each Guest that is pleasing and appealing
   - Enter the results into a database that links a network of museums, creating global routes for cities and regions according to the requests of the Visitor, with accompanying tourist routes.
   - Insert, with the cooperation of the institutions/businesses concerned, the events, the traditional paths, the chains of shows, sporting events, in the path of each Guest

- Design cultural and tourist routes for greater numbers of people, e.g., for families, couples
- Building cultural and tourist routes with a focus on local events
- Distribute educational material on local culture that enhances the routes in every country, providing the visitor with new emotional aspects
- Provide a new cultural world with a historical dimension and also hyperstorical
- Custom Software Development for events that encourage personal participation
- Personalized tips for the use of personalized courses that suit your preferences and allow you to report your favorite points
- Personal cultural profile construction with presentations
- Listing of Museums along tourist paths according to cultural demands, with complete directions.

Courtesy National Archaeological Museum of Sibaritide

Hand sign

SYMBOLOGY AND EMOTIONS OVER MILLENIA

11.4.1 Critical Aspects

1. We want this new approach of the DYNAMIC MUSEUMS FOR SOCIAL LEARNING Project to be rich and fertile. The entrance should welcome the visitor who desires culture and hopes to acquire it in the best possible way. The project offers the opportunity to evaluate behavior according to logical and emotional elements. The path forms in a hierarchical manner the birth of a hierarchy of objects, in which the hierarchy is not imposed by the person. The hierarchy is imposed by the intersection of personal data on behavior with the feedback collected on the functional characteristics of the objects, which are provided by experts and curators. They express the relationship between the "smart" objects and the human reality that surrounds the person upon visit. A list can then be created featuring numerical scores, listing the highest score first, and the lowest score last. For psychological reasons, the project displays only the first three elements, considering them as sufficient to represent a reference for the visitor.
2. Fulfillment of the visitor. The program is faced with the need to apply two logical structures: the first involves obtaining the hierarchical ranking of intelligent objects; the second must define the fulfillment of the visitor.
3. One important step is linked to the fact that not only is one museum being incorporated, as is the basis for most reference software, but rather a number of museums are being connected. This happens according to the distribution of the museums, even in very remote parts of the world. The path obtained becomes a virtual path, which is, of course, not technically real, because the museums are in different parts of the world and are associated only through the desire of an individual to visit objects with those specific characteristics. The results can be entered into a database and recognized with one click, with an image, with a code. Today, this is very easy and does not require a password. It serves only the will to know. The cultural trail also has to involve the city and the place where the smart objects live.
4. The important aspects of the project need to be developed and put in place with the cooperation of the event's organizers,
5. Paths also need to be fashioned beyond those for a single person, opening them up to more people, to groups of people. The cultural attitudes of couples in museums, and even those of friends, can be very different. We can also build trails for families.
6. We can build an effective tool for a family as a group of people with a cultural structure that is at least generally similar, to make overall management of the information on the logical and emotional behaviors easier. The introduction of data can be characteristic of a family. Cultural routes for families, couples, interest groups and even classes of students can be prepared.
7. The project is most interested in using those files that relate to the story of a person. The person must offer some guidance on the events associated with word files or images, movies or music. The association must be constructed by

the Curator of the project—and not the museum itself—who knows how to interpret the emotional reality of the person. This wealth of life can emerge as a flower that blooms out of the museum visit through the DYNAMIC MUSEUMS FOR SOCIAL LEARNING Project. The goal is a very complex one, but it is also clear that we have today, by ourselves and in collaboration with others, the ability to define the story of our lives in an emotional way. Every one of us can associate music, sounds, color, impressions, and sensations with both individual and collective experiences.

8. Another important point, however, is the creation of a distributed intelligence, evaluated for each person and capable of adding something relevant to that person. Basically, we must create a distributed intelligence related to the smart objects, modifying the theoretical cultural path that springs from behavior within a particular museum, or many museums, and evaluate the results according to the actual personality of the visitor. This program executes this initiative, absorbing the data that comes from the external environment. Visits that result in painful experiences are also important, and can hold cultural, emotional, and often secret, personal aspects for every person. This distributed intelligence must be built through new forms of expression, always now through the human interpretation of logic and emotion, because that is the nature of the human brain. The theoretical cultural path creates an avatar of the person, almost a second image. Changes in appearance are based on distributed intelligence in the outside world, which can be limited to a single country, a single city, a single region. A hierarchical structure is built among the most appreciated routes and cultures of living, based on the experience of each person.
9. All of the talk about a new cultural world is becoming a reality. Because the acquisition of information changes, it is easy to get to the bottom of each image, of each piece of music, of each work of art. You can then easily extract the one that envelops you and engages your interest the most. You can connect to the story of the daily lives of the people who lived 1000 years ago, and you can therefore develop a historical dimension to your memory of your favorite trails. You can call this idea "hyper-historic," because it is above and beyond history, a situation in which your feeling, your existence within the space visited, is determined by your emotions.

11.5 Economic Effects

The proposal for the "DYNAMIC MUSEUMS FOR SOCIAL LEARNING" Project offers interesting economic and financial effects, along with some occupational developments and jobs.

11.5.1 Economic and Financial Effects

The "DYNAMIC MUSEUMS FOR SOCIAL LEARNING" program, with its new approach to museum fruition, has the objective of increasing the number of visitors to museums, through the fostering of more intensive and meaningful visits, and thus increasing tourist flow. This will translate into demand for new professionals, creating a useful outlet for many young people, and an increase in the need for training.

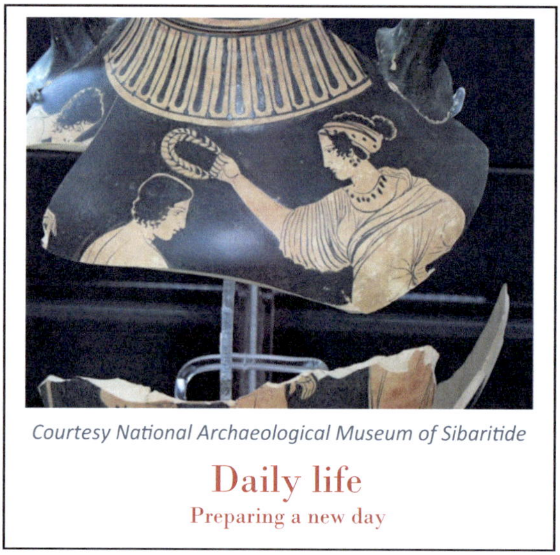

11.5.2 New Professional Figures

The program will recruit recent college graduates, or even recent high school graduates with adequate skills, who have experience in the following disciplines for the stated reasons:

(a) Disciplines related to cultural heritage, so as to develop the links between museums and citizens interested in culture.
(b) Computer sciences, for the purpose of applying and managing the "DYNAMIC MUSEUMS FOR SOCIAL LEARNING" platform, along with the data and documentation related to the various types of museum.
(c) Economics and management sciences, an area that will become especially important with the expected increase in visitors.

(d) Disciplines related to communication, so as to make contributions to the design of the reconfiguration of the museums and to play a positive supporting role for visitors.
(e) Tourism sciences, to assist in redrawing the maps of the museums, integrating them with all other events of interest for residents and tourists (movies, plays, concerts, sporting events, shows, etc.).

Courtesy National Archaeological Museum of Ferrara

Jewels

FASHION AND TIME

11.5.3 Offer of Training

The new professionalism is accompanied by an offer of educational renewal. It will be necessary that the following disciplines be integrated:

(a) Interior design, for the definition of museum exhibitions, internal distribution of the volumes, and design of the furniture and fittings.
(b) Communication design, for the creation of graphics, interactive media, signage, catalogs and all documentation, both paper and digital.
(c) Design services.
(d) Digitization of paper documents.
(e) Three-dimensional modeling.

Considering the foregoing, it would be desirable that applicants have a Degree in Museology.

11.5.4 Economic and Financial Aspects

The economic and financial aspects of the project are linked to:

(a) Establishment of digital platforms for the realization of an interconnected system
(b) Establishment of facilities to produce holograms and 3D prints for visitors
(c) Establishment of tourist facilities related to the new museum concept
(d) Development of the accommodation and catering sector
(e) Creation of tourist gadgets related to the museum, which can make for a viable promotion
(f) Internationalization of the network, with an eye towards including small museums and related network management.

11.5.5 Induced Economics

The general economic impact of the DYNAMIC MUSEUMS FOR SOCIAL LEARNING Project can be evaluated, according to the new program, in increments of 4% for high density tourist areas and 10% for low-density tourist areas. Data are taken from the development data on Museum fruition collected in Italy in 2016.

The first economic effect of the DYNAMIC MUSEUMS FOR SOCIAL LEARNING proposal is a more dynamic promotion of museums. This, in turn, will make participation of the visitor in other events in the territorial reality much smoother. The economic result is an immediate consequence of the project, insofar as it involves the visitors in a spiritual and emotional way and makes them the creators of their own visits. Today, emotional participation in events has been extended to every citizen through TV channels, the Internet, the press and the increasing involvement of communication in any given event. There is significant economic effect in the extreme mobility of the person.

The museum is a reference point, which culturally and emotionally enriches everything around it. As an example of induced economics, France, and Paris especially, always masterfully manages to occupy the top spot in the world, although the cultural richness of their museums is no greater than, for example, that of Italy.

Courtesy National Archaeological Museum of Ferrara

A Face

EXPRESSIONS AND FEELINGS

11.5.6 Economic Leadership

The DYNAMIC MUSEUMS FOR SOCIAL LEARNING Project wants to become the protagonist of a pervasive new economic reality, involving museums, the world of tourism, local events, and ultimately the participation of the visitor in the everyday lives of the places they visit. The project will prepare a map of development for employment, for new jobs needed to implement the project, requiring a global computer culture that is not yet pervasive in Italy. The map will show the types of work that will become available, related to training, documentation, and support for tourists and visitors, a version of what the Italian tourist reality already offers enhanced through the new methodology on which the DYNAMIC MUSEUMS FOR SOCIAL LEARNING Project is based.

11.5.7 Recovery of Cultural Heritage

One economic effect is the recovery of cultural objects that currently lie unseen, in stores, warehouses, archives, recently detailed in the most extensive inventories of objects. They are beautiful, but often ignored, even by the people who live on site.

This is a strong project for the purpose of getting these objects out of storage for the first time (or the first time in a long time) to be viewed by the public, not least because it can offer a dynamic structure, applicable, safe, and economically sound, for transposing their existence into one that takes place in the open light.

11.6 Action Modes

The DYNAMIC MUSEUMS FOR SOCIAL LEARNING Project does not use the usual data mining models, which combine several options by choosing the fixed data. It adopts an evaluation of each object so as to form the best suggested routes to take while visiting the Museum. It arises from the latest techniques and methods of modern artificial intelligence, including Deep Learning, which is being used in the context of museums for the first time.

11.6.1 Intervention and Planning Actions: Method

The DYNAMIC MUSEUMS FOR SOCIAL LEARNING Project radically transforms the use of museums by engaging the visitor's personality through both logical cognitive and emotional aspects, as expressed in daily life in the actions of, respectively, the left and right sides of the brain. From roughly the age of ten years, the activities of the hemisphere of emotions are constantly redesigned and developed through benefit of new knowledge.

The DYNAMIC MUSEUMS FOR SOCIAL LEARNING Project boosts the visitor's interest in convergence of the logical and emotional through participation, via computer, of the curator of the museum. You can use computers, cell phones, tablets, or any other communication device, as well as a computer station located within the Museum itself. Through mathematical algorithms, the DYNAMIC MUSEUMS FOR SOCIAL LEARNING Project generates a numerical and physical platform of suggested results for the Visitor, listing the objects most likely to be pleasing and interesting for them. The new service takes place through identification of paths between multiple museums, offering all the necessary information online about how to reach and access the museums, as well as that regarding the surrounding facilities to make for a more comfortable and appealing visit.

The DYNAMIC MUSEUMS FOR SOCIAL LEARNING Project represents significant evolution of the concept of visiting a museum, as it is traditionally understood. It heavily involves the visitor, who is transformed from a passive observer (sometimes with little actual interest) into the master of the visit.

The main points of the project are embodied in the following actions:

a. The visitor expresses his desires and interests by assigning a score from 0 to 10 to several logical and emotional characteristics. Such distribution may be accomplished by computer, mobile phone, tablet, at a fixed computer station located at the entrance to the museum or through any other type of digital information apparatus. The logical and, above all, emotional characteristics influence the mood of the subject, and therefore can vary from moment to moment.
b. The museum's curator assigns scores to the same logical and emotional characteristics of the various exhibited objects. A simple guide, prepared by the project team, directed to the Visitor, explains the definition of the Curator's scores from his point of view, and how those scores are allocated.
c. The system, through mathematical algorithms, generates a numerical-physical platform for the Visitor. The results create a customized path, cut to accommodate the mood of the moment. Each of the different pieces that make up this path can then be connected, linking all of their relative artistic, historical, and biographical details.
d. This information can then be integrated with all the local logistic and tourist information, such as, for example, walking distances, transportation, hotels, bars, restaurants and everything else that can make the visit complete, comfortable and appealing.
e. Finally, after the visit, upon leaving the museum, the visitor can realize 'something' to take home and complete the visit: texts, photos, holography, or 3D prints that document the objects that inspired the most interest, and that will enrich the memory of the visit.
f. The strengths of the project are therefore attributable to the following:

    (a) Visitor Satisfaction, the central purpose of the visit, the desired result being that the visitors come to feel almost as if the museum had been created exclusively for them;
    (b) The strong emphasis given to the emotional aspect of the exhibits, a facet that, according to the neurobiological experience of the team, is almost equal to the logical experience. Not surprisingly, the project was developed and is increasingly applied in collaboration with expert neurologists.
    (c) The consequences of a generalized application of the MUSEUMS FOR SOCIAL LEARNING DYNAMIC project, which boil down to the following:

        – The ultimate satisfaction of the Visitors, who are put at the center of the visit and "taken in hand" for the discovery of pieces of greatest interest to them. The visitor can then become a promoter of the museum to friends and acquaintances, consequently increasing the number of visitors.
        – Increased museum attendance then translates into an increase in jobs.
        – The collection of a statistically significant database to be used as feedback for the curator of the museum, very useful for purposes of future reconfiguration. Moreover, other useful information can emerge from

that database. For example, some experiences suggest that the emotional aspect has almost equal importance for visitors as the logical aspect. This can be interesting input for the Curator, who may, in setting up an exhibition, decide to place special emphasis on the emotional aspect. Finally, any visitor expressing preferences for so-called "sleeping beauties," i.e., the pieces, rarely lacking great beauty, that "sleep" in storage, may cause the curator to decide that the time to display them properly has finally arrived.

11.6.2 Team Experience

The MUSEUMS FOR SOCIAL LEARNING DYNAMIC team has so far played a part in a wide variety of activities, the main points of which are detailed below.

- After the first encouraging experiences, which made it possible to fine-tune and develop the system, the application to the Maritime Museum "Gio Bono Ferrari" in Camogli was presented to the Congress of the IFToMM (International Federation of Theories of Machines and Mechanisms), held in Vicenza in December 2016. The demonstration described the application of the project to the collection of educational models of the Mechanical Department of Politecnico di Milano.
- The same presentation was given, arousing considerable interest, to the World Congress of ICOM (International Council of Museums), held in Milan in July 2016.
- The DYNAMIC MUSEUMS FOR SOCIAL LEARNING Project was then applied to 5 archaeological museums (Altino, Ferrara, Ancona, Egnatia and Sibari) chosen by the Ministry of National Heritage, Culture and Tourism (MiBACT) in the framework of cooperation between the DYNAMIC MUSEUMS FOR SOCIAL LEARNING Project and said Ministry. This application has seen, to date, the collection of more than 1600 responses from visitors, whose assessment of the DYNAMIC MUSEUMS FOR SOCIAL LEARNING Project has been largely positive (average rating: 8.2 out of 10) and allowed for the collection of a very large amount of information of interest.

Courtesy National Archaeological Museum of Ferrara

Couple

Elegance and love

11.6.3 Application within a Region

The mode of action is to present the proposal for the DYNAMIC MUSEUMS FOR SOCIAL LEARNING Project to Museums. It basically boils down to a program on a computer platform. All data are stored and used in the DYNAMIC MUSEUMS FOR SOCIAL LEARNING program. A remote or on-site Visitor at a distance can enter the program on the computer platform, or even use an application from their phone. The visitor can express his interest, indicating his state of mind and mood. It offers a course, suggests a means of transport, provides information about restaurants and hotels, offers the most convenient times for the various visits, etc. In practice, the Visitor/Tourist has his/her day(s) organized according to his/her preferences, his/her desires, will to participate in local events. The project makes evident all museums, cultural and tourist attractions, and promotional effects that might be of interest.

The Visitor using the DYNAMIC MUSEUMS FOR SOCIAL LEARNING Project easily records the values related to his mood of the moment, for the purpose of a tourist and cultural visit. The system has saved a great deal of data and information on museums, on possible sporting and cultural events, and on conditions related to venues of interest to tourists, such as hotels. The procedure is performed through the intersection of the information given by the potential Visitor on his state of mind with the data that the system has stored and prepared with the help of the Curators of Museums and cultural and tourism operators. The result is a series of figures that represent the objects that the Visitor would prefer to visit first. The process then consists in offering a course through the museum, indicating the assorted ways to reach different destinations, pointing out hotels, restaurants and tourist attractions along the way. The preparation of a proposal for travel planning, transport, offers of tickets for the theater or football matches, representing everything of interest within the territory, turns the Visitor into the protagonist of own visit.

Courtesy National Archaeological Museum of Altino

Memories

PARENTS FRIENDS

11.7 Instant Planning

The "DYNAMIC MUSEUMS FOR SOCIAL LEARNING" Program is supported by the FOUNDATION POLITECNICO DI MILANO, as well as some Departments of the Politecnico Milan and external operating companies.

Courtesy Collection AR

Teams

OBJECTS IDEAS TABLES DEVICES
COMMUNICATION BOOKS LINES

11.7.1 Content and Structure

For development of technical and operational support for the DYNAMIC MUSEUMS FOR SOCIAL LEARNING program, beyond our current sponsorship by the Politecnico di Milano, recourse will be made to foundations and institutions interested in cultural development, Ministries and Italian cultural institutions, and foreign institutions concerned with development in Italy. Companies, individuals, and other entities interested in the promotion of culture and knowledge through Museums can furnish the economic support, combined with the active participation of the Museums. This is a necessary condition for harmonious and coherent development. The content of the project involves a modern computer system, using the most advanced technology, and a careful consideration of neurobiology, which gives us insight into the knowledge needed to fashion visits to attractions of cultural and tourist interest for both national and international visitors, who can see the remote Museum paths created for them in advance.

The project proposed involves the availability of an artificial intelligence system with innovative models capable of cataloging objects within museums, local events, and sports and celebratory occasions taking place within the territory. It is thereby able to highlight all of the human and cultural riches of the territory, placing them all within a "basket" from which the program draws the most pleasing objects, and those most preferred by users. The content, therefore, is new because it comes from a new personal and emotional place, resulting from the project's having tailored it to the emotional and intellectual preferences of each Visitor. The project makes sure that every museum object, every tourist event, every local collective expression is interpreted in terms of knowledge and emotion. It follows that the visitor becomes a very intensely active protagonist. The structure of the system rests on the foundations of computer science, neurobiology, and psychological acuity, in mutual service to the growth of the visitors' knowledge and cultivation of loyalty to the area within them. This content can be applied worldwide. We realize that this may seem like a complex undertaking. It will be simplified if we are able to involve a number as great as possible of national and international institutions in cooperation on the project. After the first phase of testing, enhancement of the project within Italian museum institutions will be ongoing.

11.7.2 Economic Forecasts and Road Maps

Economic forecasts are based on the experiences that have already been documented over a number of months within the Italian territory using a prototype of the DYNAMIC MUSEUMS FOR SOCIAL LEARNING Project. The project, from the systemic point of view, works fine, encountered no major problems in the months of experimentation within the territory, and allows for controlled development of the software. Economic forecasts are therefore related to the will to support organic growth and expansion in the area, confined, for now, to Italy. The project will continue to expand. It is currently sufficient to provide a tablet, and soon will simply involve downloading an application onto a phone or a computer. All intelligence will remain in the hands of the proponents, and the project will retain all data to the extent possible, providing evaluation according to demands and emerging needs. It is now able to expand throughout the territory and around the world at minimal cost, thanks to availability of the system.

Courtesy National Archaeological Museum of Altino

Cups

DAILY LIFE WITH ELEGANCE

Chapter 12
Project "Dynamic Museums for Social Learning"

12.1 Features of the "Dynamic Museums for Social Learning" Project

The project allows us to offer a unique new system, created within the Museums themselves, which combines the character and personality that defines a person's behavior, his/her cultural desires, and turns objects in museums into instruments of knowledge, participation, and emotion for each Visitor. The active collaboration with the Ministry of Heritage, Cultural Activities and Tourism and with the Fondazione Politecnico di Milano has allowed us to apply the project to five national archaeological museums, and publish the results. The benefits to Italian cultural life can be enlarged, implemented and made more fertile through new technologies of communication and information. We have found great partners in the participating museums, who have shown remarkable patience with the management of the results and assisted in preparation of the information in order to form a complete picture. The Curators' profound knowledge about the objects and their great love for the museum reality has been in full evidence since initiation of the project, not to mention their willingness to share and communicate with the Visitors, who, although the project is still completely new, have already delivered a positive rating on its effectiveness, over 82%, an unexpected achievement.

12.1.1 The Project

The objective of the project is to put the visitor at the center of their Museum, with the Curator acting as an interpreter of the logical and emotional reality of the "objects." The Visitor is called upon to become an active participatory subject in the visit. This is done by inviting the visitor at the beginning of the visit to express his desires, using neurobiological aspects that support both the logical and emotional

A. Rovetta and E. Rovida, *Scientific Knowledge Communication in Museums*,
Research for Development, https://doi.org/10.1007/978-3-319-68330-0_12

models, in a computer program on a digital platform, easy and fun to engage. According to the information supplied by the Curator, the program generates a personalized museum. The project was presented and applied within some very unique museums, such as the Maritime Museum of Camogli, the Natural History Museum of Milan, etc. The experiment was also demonstrated during the Convention held jointly by Milan Polytechnic Foundation and the Ministry MiBACT 5 Archaeological Museums.

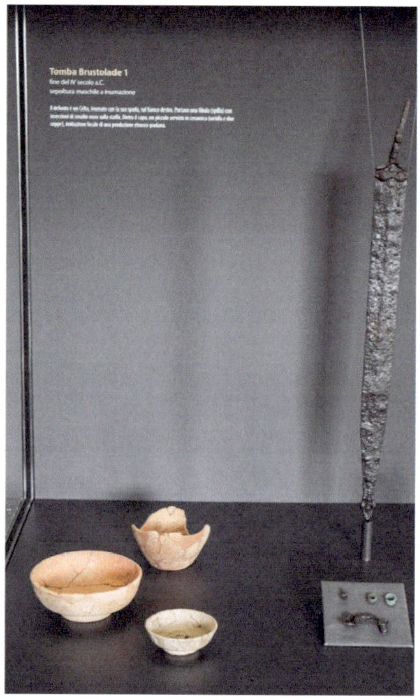

Courtesy National Archaeological Museum of Altino

Table with cups
ON THE DAY

12.1.2 Contact with Collaborating Institutions

After initial contact with the Departments of the Politecnico di Milano, we requested and obtained the support of the Fondazione Politecnico di Milano and then the MiBACT Ministry of National Heritage, Culture and Tourism, the Polihub, the Cultural Heritage Center of the Polytechnic and ICOM (International Council of Museums).

The trial run successfully created a mandate for MiBACT.

After a meeting in Rome on September 15th, 2015, to present the DYNAMIC MUSEUMS FOR SOCIAL LEARNING Project, Arch. Manuel Guido, Director of MIBACT, proposed an experiment to take place within the Italian National Archaeological Museum, which was accepted, at no cost, except for the possible Sponsor support. In talks with Minister Dario Franceschini, we made the point about the "exploitation" of the Museum property and its possible links with the tourism aspect. The development is aimed in this direction and we have arranged numerous meetings with institutions and personalities to establish a solid basis for this development. During the ICOM Conference in Milan, July 5th, 2016, Arch. Manuel Guido and Dr. Valentino Nizzo, also of MiBACT, brought the Convention required for the testing of 5 Archaeological Museums. The document was countersigned by the Fondazione Politecnico di Milano. Thus, in July 2016, the DYNAMIC MUSEUMS FOR SOCIAL LEARNING Project was given the task of conducting a developmental project within five Italian archaeological museums. It was thought that such experimentation could lead to results that would be presented in the most appropriate forum, and was not tied to a particular date for conclusion.

12.1.3 What Has Been Done

Contacts have been established with the Curators of the 5 Museums proposed by Arch. Guido and indicated by Dr. Nizzo. The museums and their curators are as follows:

1. National Archaeological Museum of Altino—Dir.: Dr. Mariolina Gamba
2. National Archaeological Museum of Ancona—Dir.: Dr. Nicoletta Frapiccini
3. National Archaeological Museum of Egnazia—Dir.: Dr. Angela Ciancio
4. National Archaeological Museum of Ferrara—Dir.: Dr. Paola Desantis
5. National Archaeological Museum of Sibari—Dir.: Dr. Angela Tecce

The figure shows a path cut according to the needs expressed by individual visitors, spurred by the recommendations offered by a guest of the National Archaeological Museum of Altino.

12.1.4 Results—Performance Indicators

Experimental measurements from 2.8.16 to 3.9.16: in that period, 1021 surveys were filled out using the Project's application.

Courtesy National Archaeological Museum of Altino

Museum	Applications	Written	Number
Altino	138	119	114
Ancona	132	61	21
Egnazia	179	167	113
Ferrara	45	0	0
Sibari	527	464	108
	1021	811	356

In the period from August 2 to November 3, 2016, 1508 experimental measurements were taken.

From the data reported in the trial, we deduced the following performance measures:

1. The average approval rating, from visitors, was 8.2 points out of 10 (scale of customer satisfaction);
2. The approval rating from the editors, evaluated by those curators who are themselves available to recommend the "game" to other curators, was 8.5 points/10;
3. The approval rating of the curators in regard to their relationship with the DYNAMIC MUSEUMS FOR SOCIAL LEARNING teams was 10/10 points;
4. The reported preference for logical characteristics was 56.5%; for emotional characteristics, it was 43.5%;
5. Visitors up to 35 years of age accounted for 33.4% of the profiles; visitors aged 35–55 and over accounted for 58%;
6. The relevant percentage has shown that males make up 49.7% of respondents, and females 42.6%, while 7.7% did not respond.

12.1.5 Substantive Proposal

Based on the results, the main substantive proposals are:

- Apply the MUSEUMS FOR SOCIAL LEARNING DYNAMIC project to a museum for the purpose of creating a personal journey for each Guest that is pleasing and appealing
- Enter the results into a large database serving a network of museums, creating global routes for cities and regions, according to the requests of the Visitor, with accompanying tourist routes
- Insert, with the cooperation of the institutions concerned, the events, the historical sites, local live entertainment and sporting events, into the path that each Guest will take
- Provide the visitor with sufficient information to take advantage of every aspect of the culture no matter where they go, with emphasis on the emotional aspects they seek
- Emphasize the historical dimensions of the attractions Listing of Museums along tourist paths according to the visitor's cultural demands, with directions included.

12.1.6 The Operational Contacts

The DYNAMIC MUSEUMS FOR SOCIAL LEARNING Team contacted the following institutions for development of the Project:

1. MiBACT, Ministry of National Heritage, Cultural Activities and Tourism for development of the Convention achieved by the Fondazione Politecnico di Milano, towards implementation of the project in Italy
2. The Polytechnic of Milan for presentation of the Project coram populo, May 19, 2016, in the Aula Magna of the Politecnico
3. The Committee of the Senate of Industry, Trade and Tourism for national presentation of the project through the main Italian institutions.
4. The City of Milan for the Milanese Museums
5. Museums of various types and Italian businesses, to show the unique and brilliant articulation of the Project in the Maritime Museum of Camogli, the Great War of San Colombano (BS), Milan Design, of Vigevano shoe, and the Archaeological Museum of Brescia
6. Cultural and economic institutions of various types, to establish connections and strengthen the possibility of applications within the Italian territory, such as the next Bank, Banca Intesa, Milanesi Museums, Museums Enterprise, the Lombardy region, the Piedmont region, the City of Florence, the City of Brescia, AIB, and many other Italian contacts

7. International contacts at the highest level, with immediate accession to the most prominent international institutions, both operational, cultural and scientific, such as:

    (a) ICOM, the World Council of Museums, July 3–9, 2016, in Milan, with open partnerships with the International Presidency and the Italian branch of ICOM, and presentation of two demonstrations at the World Congress in Milan, contacts to be maintained with the institutions for solid future international application, although only AFTER developments have been tested in Italy.
    (b) IFToMM, the Scientific Association for the Science of Mechanics, with the presentation of works to the IFToMM Congress on the issues, with scientific examples.
    (c) FAST, the Federation of Scientific and Technical Associations, with an October 2016 presentation of the project that was attended by 150 journalists.
    (d) Neurology experts who will help to make the deep neurological structure of the project, a project begun at the Eccles Foundation in 1995.

Part VI
Scientific and Cultural Trends

Chapter 13
Connecting Minds Through Networks for Knowledge and Action: Museums and Intelligent Robotics

Natural science and computer technologies are being invaded by new experiences and new knowledge about the connections between the hemispheres of the brain, with the blend of logic created by the left side and emotions interpreted by the right side. The challenge is to come to know and understand how the neural circuitry of the brain operates, how remote areas can communicate, and how the almost one hundred billion neurons and their connections can give rise to the mind, creative thinking, moods, behavior and emotions. A world cultivated through connections of information and culture, communication and participation, is a world that is enriched in its logic and its emotionality. Everyone can now be present as themselves in every place and at every time.

13.1 Neurology, Knowledge and Action

The technologies of the mind are still weak and fragile: the degeneration of the mind, seen in both the young and in the elderly in unexpected numbers and to surprising degrees, destroys any easy optimism. Today, we still do not adequately know how to deal with Alzheimer's and Parkinson's. The age of people broke through the wall of maturity, and we are required to understand interpretations of life for everyone. Neuropsychology has become immersed in the environment and in human situations, with courage and determination for a broader interpretation of emotions. Connectivity is the new basis for all study and all knowledge. The brain maps of the '90s, created by the same neurologist who developed the flight of

A. Rovetta and E. Rovida, *Scientific Knowledge Communication in Museums*, Research for Development, https://doi.org/10.1007/978-3-319-68330-0_13

supersonic aircraft with a command brain, appear now as a form of cultural archaeology. It opened the way to the contact with “smart objects.”

Chagall. For unity in mind and emotions

The inseparability of the two cerebral hemispheres has been reassessed in recent years. Knowledge is now strongly connected to the Internet, multimedia communication and computing. Data and data recollection are based on clouds and electronic devices with great capacity for storage. There are also hidden wars between powerful groups, to control the world’s knowledge and information management, the basis of any economy. All knowledge entails secretiveness, including the political and military realities of power and defense, and security needs of the future. Even for such farfetched ideas as travel to the Moon, Mars and places as yet unknown, there are calls for open minds, engineered by logic and fueled by emotions.

The cultural movements that make up our society are not only rational, but often highly emotional as well, sometimes possessed by faith and mysticism, with a desire to love things and to stay far away from worries and fears. The visual patterns, the communication on the internet presented in full color accompanied by music or sound, recall emotional and physical participation.

Also, the need to share information, knowledge and learning is reinforced, a task that becomes more and more difficult and demanding. Management of infinite data in the cloud, an information “department store” remotely accessible in mere nanoseconds, is the power of all knowledge. Data policy is linked to intelligent management of the flood of information, and the same new algorithms can also be used to manage actions of logic and emotion with a wisdom that would almost be inhuman, if they themselves were not the product of humans.

Among the symbols that interpret modern humanity (and its vision of an auspicious and positive future), there are perhaps none more representative than the knowledge collected in museums and the innovations in intelligent robotics.

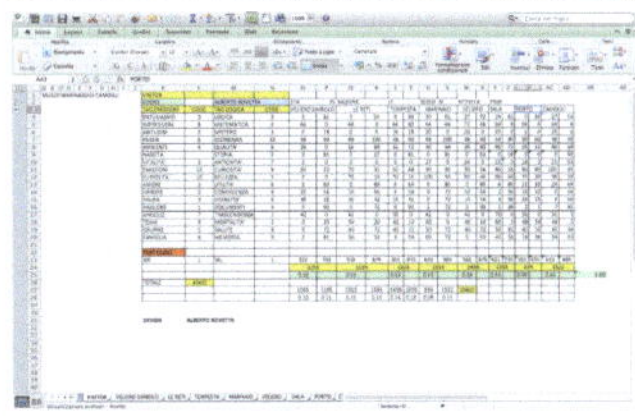

Neural networks and algorithms action

They are expressions of the many-armed monster that involves all our thoughts, all our actions, even those undertaken subconsciously.

13.2 Knowledge: The Dynamic Museum

Knowledge over the centuries was transmitted through books and stored in libraries. From the birth of the digital, knowledge has been stored in powerful digital systems with limited space and the capacity for fast and easy access anywhere in the world. The culture expressed by such significant objects as paintings, statues and archives is preserved in museums. In recent times, visitation to museums has grown by high percentages. Museums represent the expression of knowledge: we consider museums to be the first protagonists of a new acquaintance with knowledge, something a computer network cannot provide on its own. Museums are a domain of great importance, one that is growing day by day as its users increase from every social and economic sector. They are dedicated, by tradition, to the collection of objects. Today, they require intelligent objects, which must be visible, readable, understandable, acceptable, lovable, and possibly even approachable (in terms of an interactive experience). Will Museums now become properly "smart," that is, will each "object" become suitable for understanding the logic and emotional participation they may engender? Human motivation to attend a museum is linked to knowledge, empathy, and emotional limits, the very dangerous outlier of which is referred to as the "Stendhal Syndrome," an experience that leaves the visitor with an unpleasant sense of inadequacy. The most significant level of acceptability is linked to personal passion, the satisfaction, the excitement of being in front of quiet beauty, curiosity satisfied, emotional memory established. The two hemispheres of the brain work in continuous collaboration and empathic communion during a visit to a museum.

Museum collections consist of donations, of individual works or series of objects within a themed set, on any subject, reflecting personal experiences or historical events, and can be related to History, the Arts, Industry, Science, Fiction, Nature, etc.

The organization of museums is very variable: physical paths of objects with placards containing physical descriptions, mixed systems, virtual tours, itineraries along digital paths incorporating multimedia, knowledge services and selective displays, customized, digital, etc. In addition to the catalogues typically available in the Museum, the choice of information for museum sites can take place on the Internet, through search engines like Google, Baidu, and others. It is necessary and vital to have a link to the museum and a selection with all linkable Museums. The new rule is that: “You exist if you are linked; if you are not linked, you do not exist.” Sometimes it looks true. The quality of information opens up issues of value and significance. The ranking of the information can be connected to criteria of truth, quality, completeness, presentation, from which text and images and video and links then emerge and merge. Even the organization of classification is essential. For objects collected in a museum, there is always news that requires monitoring and verification, including a projection of what significance the object may have in the future. All museums are built upon basic knowledge, which must be completed through the use of logical data, and enhanced with a share of the emotional, which keeps it fresh and alive.

In the museum, the protagonist is the person, who follows his conscience, who is informed by connections to the outside.

Today, museums can follow the evolution of communication, presentation, shared participation. They rely on the left hemisphere of logic and constructive thinking, and the right hemisphere of emotion and reasoned action. The request for knowledge originates from the person, in a gesture of creative curiosity, which is vital for knowledge. From the total combined space of objects in Museums, a network of museums can be constructed, representing both profit and value.

Computer science is a rich and powerful discipline, and offers the possibility of inviting a second opinion (as in medicine advisories), that is, a plurality of opinions, even those that are discordant. Ultimately, they are all directed towards the satisfaction of the visitor’s curiosity and encouragement of their participation in an experience in which they become a tourist within their own thoughts and within their own soul.

Knowledge and modes of thinking and living have all changed over the past decade, in some ways to stunning degrees, as the world is increasingly stretched and relaxed between the past and the future. The same network updates the “container” of the person in an automatic and conscious way through icons and means of multimedia. Everyone is looking towards the morning of tomorrow, to know and to act in a life that increasingly calls for presence, participation and emotional logic. **That requires consciousness of being open to knowledge from the outside, in respect to the construction of a planet increasingly rich and purified of darker forces that operate without logic and without positive emotion.**

13.3 Action: Intelligent Robotics

Human action is so complex that the idea that thinking machines and artificial minds could possibly be a substitute for it is a wild and irreverent thought. That irreverence can become respectful scientific curiosity when we seek results that are more human, painless, peaceful, sparing the hard work and making it pleasant instead, transforming a risky business into a curious action at a distance that a robot can handle. Robots are an evolving species, timeless and with unpredictable limits, as unpredictable as those of the future of the mind and human nature (See Aristotle 330 B.C.; Eccles 1964, 1966; Edelman 1987; Levi-Montalcini 2004; Rovetta 1987; Sherrington 1906; Sperry 1970).

The concept of robotics requires a closed circuit, called a loop, in which an action causes an effect, and the effect returns into the circuit as the cause of a subsequent action. The robotics industry has thus conquered the world of work.

The sub-discipline of bio-robotics was born from the connection between creative minds and human physiological limbs, organs and the whole body. Here, the loop complex contains bodily elements, joined by mechatronic systems that are chemical, neural and physiological.

A mechanism capable of copying the actions of a human has been a constant in human aspirations and the collective imagination and fantasies of almost every civilization: the humanoid robot. A real functioning humanoid robot capable of motion was created by Professor Ichiro Kato in 1976 at Waseda University, Japan. Since then, the challenge of machines + brains, to get to mechanisms + minds, and build a system that could be considered an "intelligent robot," has been part of the progressive March of technology, science, fantasy and emotion. They reach into the curiosity of every human imagination, even conjuring up systems that simulate the external actions of animals, insects, children, and monsters. The connection of thought and emotion has also created systems that overflow beyond the mere copying of the natural, inventing and realizing the creation of new functions (such as the space robot Ladyfly, for which new movements that do not even exist in nature were invented). In recent years, the power of communication, information technology, and knowledge management has allowed for the building of robots that function as natural systems: "mechanical logic + emotion."

The new knowledge related to neurology and neuropsychology is slowly pervading science and technology today, firing bullets into the reality of the human brain. Emotions further infiltrate robotics on a daily basis.[1,2]

After creation of a single "intelligent" robot, the next powerful step for the future would be a colony of robots, either programmed to function in a hierarchical setting, or in mutual cooperation. The most significant step would be to define what an intelligent and emotional robot is. It must be equipped with circuits and sensors that translate logical expressions and must experience sensations of the human type: sensations such

[1]http://robotica.mecc.polimi.it

[2]http://www.locobot.eu

as vision, touch, taste, smell, a sense of the temperature, the humidity, and even more, indeed, whatever the technology allows. Today, robotics can work alongside human action, or it can cancel out the human effort, relying on control mechanisms that are flexible, intelligent and conscious. Thus, robotics slips easily into our daily activities.

Robotics could offer a mathematical model for the dynamic behavior of the machine, which is the result of a software program developed by human operators. If you could identify and separate the different elements of electronic circuits and software components that oversee the machine's movement and perceptions, it would be useful for the technologist and for biological simulation experts. If a bridge could be built between robotics and human neurological reality, then intelligent robotics could become a field of application and evaluation of theories and knowledge about the human brain and its behavior.

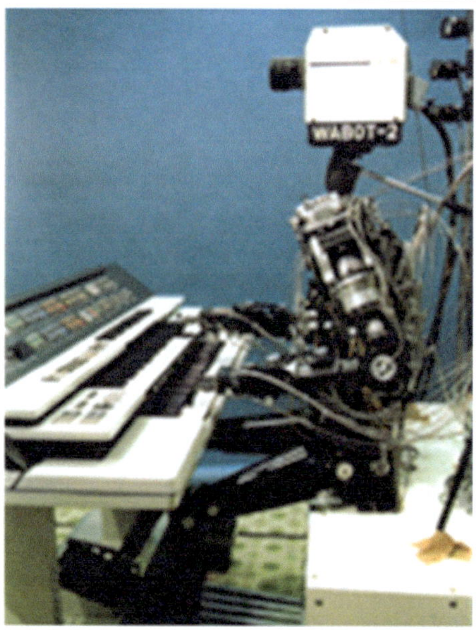

Courtesy Collection AR

Prof. Ichiro Kato - Pianist robot

With updated neuropsychological participation, even those who work on computers could contribute through shared museums and intelligent robotics. With science, will, faith, and the courage to brave the unknown, we will be prepared to engage in a participatory adaptation of our present, one that is increasingly rich and challenging.

References

Aristotle, (330 B.C.), Categories, 8; 9

Eccles JC (1964) The physiology of synapses. Springer, Gottingen

Eccles JC (1966) Brain and conscious experience. Springer, Heidelberg

Edelman GM (1987) Neural darwinism, New York

Levi-Montalcini R (2004) Abbi il coraggio di conoscere, Milan

Rovetta A (1987) Voice control. In: Encyclopedia of robotics, vol III, Ed. Dorf

Sherrington CS (1906) The integrative action of the nervous system. Yale University Press, New Haven, London

Sperry RW (1970) Cerebral dominance in perception, in early experience in visual information processing. National Academy of Sciences, Washington

Chapter 14
From Brains to Cybernetics

New knowledge about the mind has resulted in new interpretations regarding the nature of the relationship between the brain and the mind. Technologically advanced systems (such as robots and computers) created by man to copy certain functions of the human body and mind, despite failing in the more difficult tasks, have nonetheless succeeded in giving way to new outlooks regarding awareness of the way the human mind works.

Current studies are operating based on evaluation of the results obtained so far from robots and computers effecting simulations of the human mind and body. Although reconstruction of the human brain and body using current information and mechatronic technologies originally appeared to be a pointless enterprise, it has now proved to be a critical basis for the reproduction of some specifically human functions.

We are now entering a world in which extra-cranial knowledge is creating, through a process of evolution, brains adapted to a world simulated by means of modelling, which can be just as aggressive and just as appealing.

Natural science and computer technologies are both being informed by new experiences and knowledge of the connections between the hemispheres of the brain, with the blend of logic created by the left side and emotions interpreted by the right side. The results of optogenetics and new applications of fMRI, functional magnetic resonance imaging, are being carefully studied. The challenge is to come to know and understand how the neural circuitry of the brain operates, how remote areas can communicate and how the almost one hundred billion neurons and their connections can give rise to the mind, with its creative thinking, moods, behavior and emotions.

Robotics could offer a mathematical model of the dynamic behavior of the machine, which is the result of a software program developed by human operators. If you could identify and separate the different elements of electronic circuits and software components that oversee movement and perceptions, it would be useful for the technologist and for biological simulation experts. If a bridge could be built between robotics and human neurological reality, then intelligent robotics could

A. Rovetta and E. Rovida, *Scientific Knowledge Communication in Museums*,
Research for Development, https://doi.org/10.1007/978-3-319-68330-0_14

become a field of application and evaluation of theories and knowledge about the human brain and its behavior.

The results outlined in this book show how cerebral and neuromotor activity responses to physical input come from the real world.

14.1 Symmetry and Hyper-symmetry Between Cybernetics and Brain Science

The computer operates under serial sequential conditions. Only by imposing a certain parallelism between parts of the computer programs can a certain symmetry inherent in parallelism be reproduced. Instead, we see a centrality of behavior in the computer, which receives digitalized signals and converts them into signs for machine control. Geometric symmetries rarely exist in the software, which uses loops, logic comparisons and controlled loops. On the other hand, there is a specular symmetry between the software program and the action of the robot, which physically executes the logic structure of the command program.

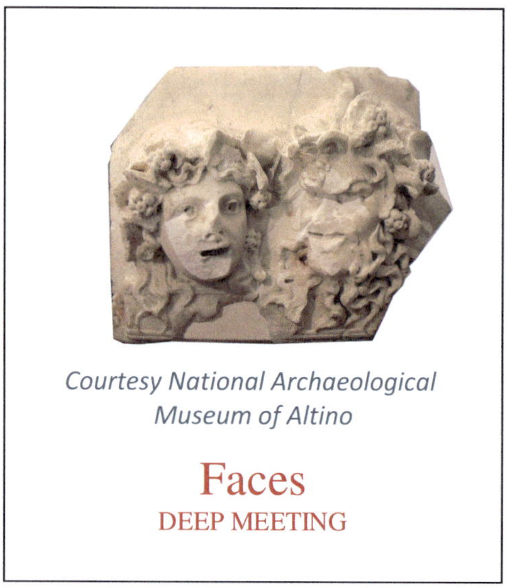

The scientific approach must be stronger: the symmetry of cybernetics in relation to the mind + the human brain does exist if we consider that both realities originate from the same trunk, around a tree.

The tree-type logic structure pervades the discipline called “artificial intelligence,” but the real connection between cybernetics and brain science requires a three-dimensional view of the symmetry structure.

Symmetry is a spatial symmetry, i.e., a special hyper-symmetry that we are transcending on two fronts, one on the part of neurobiology, and the other on the part of cybernetics. Today, to forecast and understand the physics of things, the brain uses mathematics to create computational models and then uses those models to simulate images and sensations, thus giving an impression of something that exists outside of the program.

For the moment, the results of modelization appear to be rather positive. However, these are linked to the evolution of human culture and will continue to grow with it. Cybernetics has already used mathematical models and modelling, with disappointing results. Nevertheless, the study and application of mathematical models and modelled environments has shown a scientific potential that has only been partially exploited until now, and will become more and more positive and efficient as the work progresses. The mathematical model of "direct + inverse" kinematics has only been in existence for a few years, and is, in any case, the one instrument capable of predicting the dynamic movement of a system today.

Those active in the field of neurosciences should observe the theoretical and practical results obtained by cybernetic experts, so that the vast quantity of work can be simplified into some solid theoretical bases, capable of transferring results between the two groups of scientific disciplines.

The general structure of the brain, whose method of operation is still to be discovered, requires the cooperation of all of the most important neuroscientists, who need to learn to look at things from a different viewpoint, to obtain practical results, by identifying the roots of knowledge through the results obtained by cybernetics in regard to the science of the brain and of the mind.

14.2 Expectations from Cybernetics and Neurosciences

Over the last few years, those involved in robotics, bio-robotics and cybernetics have been looking forward to some new contributions in the fields of neurobiology and cognitive science that will allow for advances towards more "human" machines. Three aspects of machines are still very weak: intelligence, sensitivity and awareness. This means that the dream of a "human" machine is still remote. Still, the symmetries that the human body and the cybernetic robot seem to offer are apparent.

Courtesy National Archaeological Museum of Altino

Bottle of glass
BEGINNING OF VENICE GLASSES

These are "hyper symmetries" that transfer some basic aspects of the human organism (human) to other aspects of the cybernetic machine (the robot). Although there are functional symmetries within these two entities (human and machine), until now, little has been done to establish genuine analogies (Basmajran 1971; Changeux 1983).

The effort involved in constructing a copy of the brain and the body using current information technologies and mechatronics has proved to be pointless. Scientific results, which are purely technological and deductive, have been expected from mechatronic methodologies (Eccles 1994). Other disciplines, ranging from neurosciences to robotics, from logic to computer sciences, have proposed new approaches aimed at interpreting the behavior of the human brain.

A brief analysis of the control within "brains" of the most modem cybernetic machines shows how current robot structures, for example, do not offer any aspect of mental creativity (Dorf 1989). At the same time, over the last few years, the modelling of external reality has introduced, into the most industrialized countries with high communication levels, new real and virtual machines, with graphic images. Day by day, within the context of everyday reality, we see the cerebral, mental and psychic participation of healthy people in digitalized perception and communication processes that are completely new to human nature (Edelman 1989; Gregory 1987; Llinas 1988).

Reality modelled and simulated by means of computers and graphic systems can become, for some people, more reliable than reality itself, when mental abandon and perceptive participation reach certain levels of confidence.

In the face of simulated reality, the reactive human response, in healthy people, appears to be much quicker and even more immediate than the response to actual reality; neuromotor response times are shorter. We are entering a world in which extra-cranial knowledge is, through evolution, creating brains suited to a simulated world, which is just as aggressive and appealing as the real world (Penrose 1989).

Fantasy is pervasively dissipating the conditions of attachment and adherence to reality (Edelman 1989).

14.3 Human Brain and Robot Brain

The anatomy and physiology of both human and animal brains have been generally described in numerous texts.

The brain of the robot is called a "brain" because it is maintained that the processing of programs, which are translated into motor actions of the robotic machine, take place within it.

An examination of the functioning of a robot "brain," with a sample of approximately two million robots, shows how nothing that can genuinely be considered cerebral takes place within it. Some recently applied methodologies represent developments in algorithms, mathematical genetic structures and parallel computation systems (Dorf 1989).

The natural sciences and computer technologies are being informed by new experiences and knowledge of the connections between the hemispheres of the brain, with the blend of logic created by the left side and emotions interpreted by the right side. The results of optogenetics and new applications of fMRI, functional magnetic resonance imaging, are being carefully studied. The challenge is to come to know and understand how the neural circuitry of the brain operates, how remote areas can communicate and how the almost one hundred billion neurons and their connections can give rise to the mind, with its creative thinking, moods, behavior and emotions[1] (Richards 1988; Searle 1980).

Robotics could offer a mathematical model of the dynamic behavior of the machine, which is the result of a software program developed by human operators. If you could identify and separate the different elements of electronic circuits and software components that oversee movement and perceptions, it would be useful for the technologist and for biological simulation experts. If a bridge could be built between robotics and human neurological reality, then intelligent robotics could become a field of application and evaluation of theories and knowledge about the human brain and its behavior.

[1]http://robotica.mecc.polimi.it

14.4 Cybernetics for the "Intelligent" Robot Project

The way in which a robot behaves depends entirely on the designer's decision to make it perform tasks that are programmed per a rational logic scheme. As far as the robot is concerned, there are no functions such as "enthusiasm," "conviction," "doubt," etc., but only mechanical engineering operations devoid of intelligence or sensitivity.

More advanced tasks that simulate robot sensitivity are simply a communicative representation of attitudes and appearances that give those watching the idea that the robot's brain is causing it to adopt human attitudes.

The robotic machine and its "computer" component possess neither intelligence nor sensitivity, other than those induced by the operator. The operator can make the robot cry if he wishes, through use of a simple hydraulic drop circuit. The robot cannot, of course, cry by itself, and although this seems like an obvious observation, this reality is often surprisingly difficult to convey, even to people of a certain cultural background.

14.5 The Human Brain in Robot Programming

The programming of a robot and its behavior is completely neutral from a cerebral point of view. The designer of robot control can limit herself to drawing a flow chart of movement. Any good programmer uses the language that she knows best and transforms the flow chart into a program with instructions that are both legible and easy to understand. The robot absorbs these instructions into its memory and faithfully performs them, within the limits of its characteristics.

The programming of behavior does not foresee any reaction of a cerebral or reactive nature, other than that programmed per external feedback. It results from measuring instruments (called sensors) that impose feedback on movement or behavior in general.

14.6 Senses, Brain and Awareness in Cybernetics

A robot's "brain" is constituted by its electronic control, which handles and operates the movement of the robotic arms and the overall behavior of the machine simultaneously. The robot sensors are measurement instruments that only read the physical amplitudes for which they have been designed—a way of acquiring an engineering-type sensory linked to the shape of the sensors. A force sensor only detects the value of a force, and not that of a temperature or pressure.

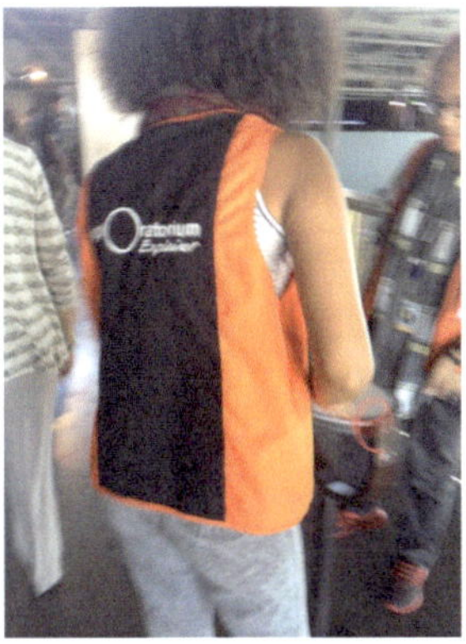

Courtesy Collection AR

Learning and teaching

ALONG THE ROADS OF KNOWLEDGE AND EXPERIENCE

The human body is able to perceive hundreds of amplitudes at the same time, with millions of simultaneous signals that will be then processed by the brain.

The same does not apply to the robot, which carries out sequential analyses that, if parallel, are sequential in their logic.

The participation of a robot in an environmental system is, as things presently stand, just a matter of semantics. Participation only means that the machine is physically present in the system and interacts with it. The "conscience" of a robot, like the conscience of a machine, an electronic control system or a series of circuits for programming, is non-existent.

Auto-programming of a robot, if present, is an intellectual game played by the programmer, who elevates the tone of automation and permits the robot to follow several possible behavioral patterns. The choice of the final pattern is conditioned by the option instructions, designed and prepared by the control designer.

14.7 Animal Circuits and Cybernetic and Robotic Circuits

If a robot were to have feedback and feedforward circuits, its behavior, in a certain sense, would be similar to human behavior. Analysis of the computational software program of a robot shows that its actions during each instant of motion are not only fixed in terms of time and duration, but also predictable. In a human, each action is, per se, alternative and variable, in relation to external sensory factors and internal decisions.

What are these channels of reasoning and action? How do they interact? What are the reciprocal response times in the face of a similar task? How can the tasks of a robot be compared to those of a cerebral human?

Will the future of the robot in the history of cybernetic technology be like that of a cerebral human, with minimal corporeity, or to that of a physical human, without a brain and a high degree of electronic control?

References

Basmajran JV (1971) Muscles alive. Williams and Wilkins, Baltimore
Changeux JP (1983) L'homme neuronal. Fayard, Paris
Dorf RC (ed) (1989) International encyclopedia of robotics. Wiley, New York
Eccles JC (1994) How the self controls its brain. Springer, Berlin
Edelman GM (1989) The remembered present. Basic Books, New York
Gregory RL (1987) The Oxford companion to the mind. Oxford University Press, Oxford
Llinas RR (1988) The biology of the brain, from neurons to networks, Scientific American
Penrose R (1989) The Emperor's new mind. Oxford University Press, Oxford
Richards W (1988) Natural computation. Mit Press, Cambridge, MA
Searle J (1980) The behavioral and brain sciences. Cambridge University Press, Cambridge

Chapter 15
3354 Years Ago

In speaking of museums, we seem to recall only the past. As if the past were not part of our present and as if we ourselves did not live in our own past, as we look to the future.

A. Rovetta and E. Rovida, *Scientific Knowledge Communication in Museums*,
Research for Development, https://doi.org/10.1007/978-3-319-68330-0_15

1014 A. Rovetta et al. / Mechanism and Machine Theory 35 (2000) 1013–1031

Fig. 1. A chariot belonging to the Egyptian Pharaoh Tut Ankh Amun, 1337 B.C. and a portrayal at the Egyptian Museum of Cairo (courtesy).

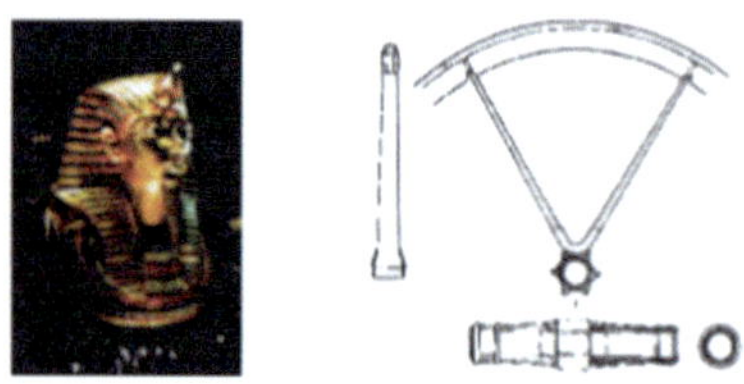

Fig. 6. The Egyptian Pharaoh Tut Ankh Amun on display at the Egyptian Museum of Cairo (courtesy) and chariots principles (see Ref. [2] Plate LV, Chariot A 5).

Pharaoh Tut Anck Amun

Chariots

SCIENCE AND TECHNOLOGY OVER MILLENIUMS

If we truly want to see the past in museums, Museums should become the forge of the future. One of the facts that has most recently impressed the imagination, consciousness, and wisdom of scientists and historians is the discovery by the Cairo Museum that the Pharaoh Tut Anck Amun's Treasure contains five chariots, the technological structure of which was recently confirmed as being definitively scientific. In the shadow of the Sphinx, the Egyptians constructed chariots bearing the same elegance and intelligence as that seen in the initial ideas for the design of cars at the end of the 1800s, 3300 years later. In the imagination, the sphinx is a mysterious entity that, according to the Egyptians, watches over life's mysteries. Along with this aspect of Ancient Egyptian culture, there were also factories and unexpected technologies, the chariots of 1337 BC being a perfect example. The wheel, the simple and universal wheel, invented by humans so as to be able to move better, became a jewel in the hands of the scientists and technicians of Egypt in 1330 BC. The figure shows a wheel, the inner part of which, the hub, is very stiff and hard, while the outer shell is very tender. How do these compare to today's bushings, evolved over the years, but applied in the 19th century? The suspension may be smart, and the wheels can be flexible and adaptable to the ground. How can we not consider these chariots, so simple, natural, and elegant, as a great source of ideas, thoughts, emotions, logic and technique? Happy cultures who know how to rest their feet in the past stay in the present by focusing their eyes on the future, remaining perfectly consistent. The wheels are pieces of shaped wood treated for almost six months, which flex with a movement that mirrors current modern intelligent suspensions. The third essential element is almost unbelievable: the outer rim of the wheel has almost the same shape as a tire, and covers the entire wheel, becoming a rolling element that favors motion. Only around the year 1850 did European and world engineers conceive the idea of using tires to complete automobiles.

A museum is not a collection of the past if it is not also a mine for the present, and for the ideas and creativity that will take us down the emotionally unexpected and unpredictable paths of the future.

15.1 The Chariots

The six chariots belonging to the Egyptian Pharaoh Tut Ankh Amun are distinguished by wheels that are both flexible and adaptable, thanks to the shape of the rim and the wheel spokes; the bearings are built by exploiting the principle of a hard material against a soft material and by applying animal grease between the surfaces. Furthermore, the long nave presents a specific low dynamic pressure, which reduces wear and tear and allows for good stability of the wheel. The position pertaining to the center of gravity of the cart and the driver is such that the dynamic moment is reduced. The horses must exert traction power with low vertical and horizontal components to drag carts along all types of ground. The functionalism and mechanical efficiency of the chariots could be the result of a lucky series of technological coincidences (flexible wheels, compliant spokes, long hubs, friction surfaces between a hard body and a soft body to favor dynamic coupling and duration and center of gravity position close to the axles). The wheels are light, elastic, and move well against the ground. This is the first conception of a tire (i.e., an element that can be adapted to soil irregularities), initiated at a time when the science of mechanics and the study of vibrations and structure dynamics had no written basis, from either a technological or theoretical point of view. Nevertheless, such performance could be justified by the scientific and technical results as being studies of the science of mechanics, even if theoretical studies on mechanics and friction did not assume an organized structure using scientific methods until the

19th Century. The wonderful coincidence of modern technology found in the chariots belonging to the Egyptian Pharaoh Tut Ankh Amun can, however, lead us to believe that moments of high creativity and elevated spirituality can produce wonderful results. Outstanding technical results, before the millennia, could be the result of spiritual freedom combined with the intelligence of a highly cultural environment.

Fig. 3 A wheel from Pharaoh Tut Ankh Amun's chariot (1337 BC) on display at the Egyptian museum in Cairo (courtesy)

Wheels
BEARINGS, SUSPENSION, TIRE
IN ANCIENT AEGYPT

In conclusion, some of the wheel kinematics and dynamics used in the 1337 B.C. chariots belonging to the Egyptian Pharaoh Tut Ankh Amun are extraordinary, insofar as they closely resemble current mechanical concepts. From a mechanical point of view, the chariots that belonged to Egyptian Pharaohs, adaptations of chariots imported by the Hittites, were very efficient and smooth-running. The chariots were discovered in 1922, in the Tomb of Pharaoh Tut Ankh Amun, together with his treasure. Completely disassembled, they were found in a corner of the tomb and were subsequently reassembled. The ingenious design, as well as the architectural and engineering skill applied to the construction of this type of chariot, is discussed in this paper. An analysis of the materials was conducted at the Egyptian Museum in Cairo, as was the reconstruction and restoration of the chariots. Some mechanical aspects of the wheel structure, spokes and materials used for the sleeve bearings are extraordinary, as is the use of animal grease as lubrication and, indeed, the entire actual design of the chariot itself. This is especially true in as much as the concepts at work are very much like current concepts. It may be concluded that high levels of cultural, spiritual and intellectual advancement favor the development of technological solutions that result in an improved quality of life. The main idea is that, to evolve, humankind, if stimulated by a high degree of cultural and spiritual energy, will invent suitable technologies, some of which will prevail, while others disappear. If the level of civilization is high, then creativity and ingeniousness develop technologies in line with the quality of life. In conclusion, these chariots appear to be the first mechanical systems involving the use of kinematics, dynamics and lubrication principles. These chariots are displayed at the Egyptian Museum in Cairo, and may be cited as examples of construction principles in transportation, well ahead of latter-day mechanical principles and laws that would only be discovered more than 3300 years later.

15.2 Different Conception

The design of the "DYNAMIC MUSEUMS FOR SOCIAL LEARNING" Project is a different concept from that usually seen. We see the reality of the culture, hear the symphony of criticism about knowledge, life, everyday life, the way humans live. The museum is a path representing the existence of things, those of the present and those of the past, any of which can be the basis for the future that begins tomorrow morning. Let us review this concept of things that belong to us and those that command us, the objects that we love and that we are required to live with, to love, to learn about. We see in the very word "museum" a representation of everything that unites people, in memories, in experience, offering courses ready to be explored. We sink into memory in the oldest museums, including those with prehistoric objects that recall the nature of the past. The dynamic museum reminds us of the frailty of human nature, with its great power. **Museums make us feel unique, featuring characters of indestructible humanity that grows. This culture of being is not only logical, it is not only about the history of the facts of living objects: it is just the death of the things which make us suffer**. Space and time become expanded, and the museum becomes the protagonist. With increasing humanity, we grow and move forward, we feel the need to know, to remember the past, not to relive it, but to allow ourselves access to the memory of sweetness and depth. That is why the new tools for information, the communication of new computer logic, and artificial intelligence can be beautiful gifts, delivered with ease and joy. Even the seriousness of the knowledge and wisdom enabled by the new technologies and the virtual representation of objects helps those who want to participate in the reality of the past, to consider the future. DYNAMIC MUSEUMS FOR SOCIAL LEARNING represents a way to effectively face the reality of all the museums of the world, to participate, to feel like protagonists, to recognize, with the help of museum curators, points of interest and participation, of human participation.

Courtesy Collection AR

Garden, Green, Building, Art

A LINK TO SPACE OF MIND ON KNOWING AND FEELING

By using appropriate data in a knowledgeable way and adopting it to the world of computers, you come to recognize the new supportive approach to knowledge.

Chapter 16
ICOM Official Presentation

For some years, the intelligence of objects has been the discipline that explains the internal aspects of the things we use in daily life. Certainly, this must be a notion worth applying to the presentation of objects in Museums, and perhaps also one that can help us find the best way to present them specifically as objects of love.

16.1 Museums and the Intelligence of Objects

Knowledge of objects draws out their intelligence, and now human knowledge has passed the external description of things and objects in general, to make them a share of the knowledge. Towards the near future, which is directed to "connecting minds." Facts have been overwhelmed by the immediacy of communication, and memories are generated within the cloud and through social networks, with billions of people—not just technology experts—using it spasmodically and in a valuable way.

16.2 Perspectives on Museums

Today, on the verge of 2018, a visit to the Museum still involves walking past objects, described on placards or through short films, requiring personal interpretation that is often unclear and not always complete, to make and create true "knowledge." The desired road, already being talked about by some people, is to create a new system that benefits the Museum and creates an Environment of Knowledge.

The Museum, like a house full of smart objects, should have no gates, no strict obligations, with the capacity for its visitors to have their eyes open to the sky, to

A. Rovetta and E. Rovida, *Scientific Knowledge Communication in Museums*,
Research for Development, https://doi.org/10.1007/978-3-319-68330-0_16

the stars. The knowledge required is easy and simple, particularly in the context of the immense capacity of human knowledge, if guided by intelligence and emotion. The IoT (Intelligence of things) is also the sum of logic and emotion, because everything is loved and understood, taught with passion.

To preserve artistic heritage is also to keep alive the resources useful for tourist activities, thus increasing not only the entrance tickets that will be torn, but also multiplying the revenue related to the surrounding areas, such as transport, catering, accommodation, communication and merchandising. Many foreigners visit Italy in an almost systematic way, going to places of personal interest that are well-known. However, much—much too much—of Italian culture and its museums is not known by foreigners, or even Italians, for that matter. We decided to take a new road that would make suggestions to museums throughout Italy as to how to improve their relationship with their visitors. This would involve talking to many different tourists: Italian, foreign, educated or those simply interested in beauty and artistic technique. They could suggest objects and routes, information and knowledge that can make their visit to the museum even more exciting. And they could use mathematical programs and neurological knowledge to serve the visitors' interests and cultural needs, both the logical and the emotional. Digital technology is an integral part of the DYNAMIC MUSEUMS FOR SOCIAL LEARNING Project and allows for the sharing of information and feelings. PCs, tablets and smartphones, connected to the network, allow for the catchment area of our application to be expanded considerably.

The DYNAMIC MUSEUMS FOR SOCIAL LEARNING Project radically transforms the use of Museums, engaging the visitors' personalities through both their logical cognitive aspects and their emotions, expressed in daily life through the actions, respectively, of the left and right brains. The DYNAMIC MUSEUMS FOR SOCIAL LEARNING Project boosts the convergence of logical and emotional interests of the visitor through participation, by computer, of the curator of the museum. This can be done through computers, cell phones, tablets, or any other communication device, as well as at a computer station located within the Museum itself. Through mathematical algorithms, the DYNAMIC MUSEUMS FOR SOCIAL LEARNING Project generates a numerical-physical platform that suggests results to the visitor, listing the most pleasant and interesting museum objects for the visitor in all the participating museums. The platform can also be useful for the reconfiguration of Museums themselves and for reconfiguration of the museum's routes, which can be prepared for each individual visitor. The new service offered by the platform also involves identification of paths connecting multiple museums, offering online access to all the information necessary to reach and access the museums, along with all the adjacent facilities that will make the visit more comfortable and appealing. Each visitor can also look forward to being able to produce printed photos, holographic images, and 3D models at the end of the visit, take-home souvenirs of those museum objects that touched them most.

Now, the process is very simple: the visitor willingly fills out a survey on their mobile phone, tablet, computer, or even on a printed sheet, answering questions about his intellectual and cultural interests, and his emotional involvement and satisfaction regarding museum objects in general. The "DYNAMIC MUSEUMS FOR SOCIAL LEARNING" Program, in turn, has data collected from Italian museums on the intellectual, cultural, and emotional value assigned to a wide variety of objects by previous visitors. The intersection of these two batches of intellectual and emotional data that have been collected results in a ranking of favorite items. These rankings can then be viewed by the visitor on their electronic devices as they decide upon their visitation intentions, accompanied by recommended routes, and with the option to receive recommendations of tourist attractions and cultural routes based on their preferences (Fig. 16.1).

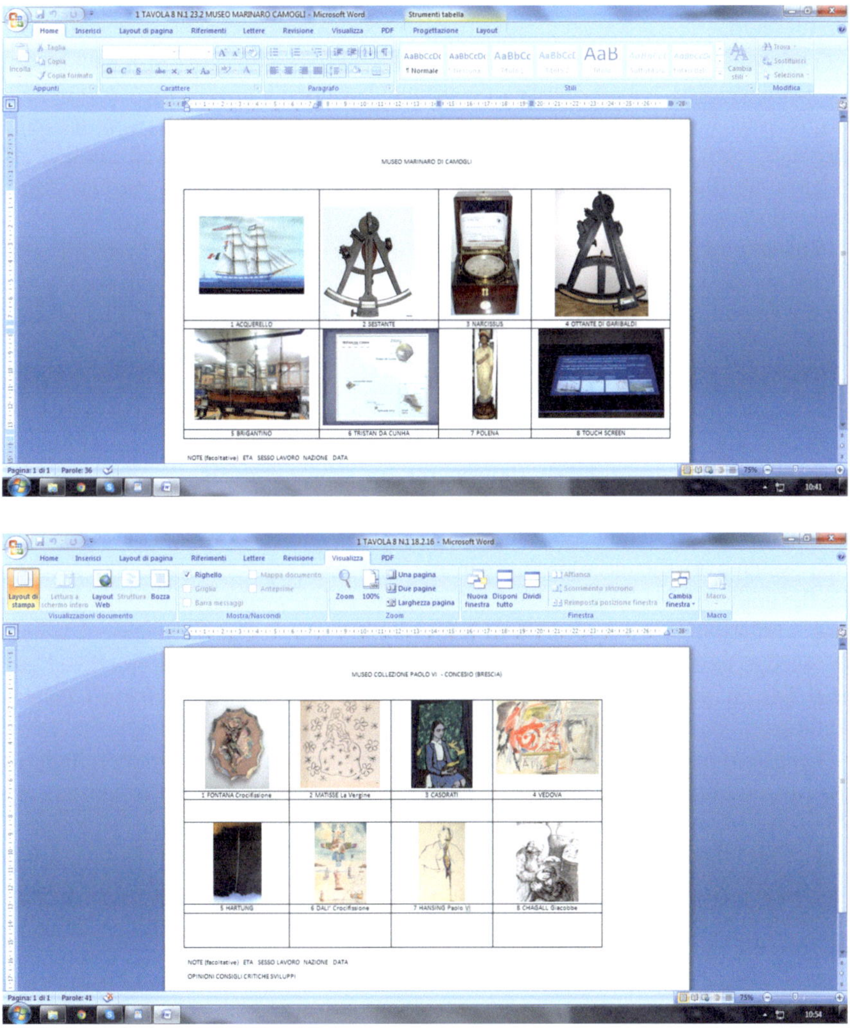

Fig. 16.1 Objects of Museo Marinaro di Camogli and Museo Paolo VI

16.3 Applications to Museums

The application possibilities for Museums are numerous. Let us look at two of these. There is an application dedicated to the Maritime Museum of Camogli, which collects objects and documents relating to the seafaring history of that Ligurian town, both regarding its great tradition of the past and its intense activity in the present. An important first signal of the application's success was in the early responses from Visitors following their visits to the Museum, a success that the museum then compounded in its collection of two further objects, one historical, the other futuristic: the octant that belonged to Giuseppe Garibaldi, and a touchscreen, which provides access to digital information and simulations of intelligent Items in the museum.

The visitors who are interested in maritime history connected to Italian history are strongly attracted to Garibaldi's octant, because it recalls the exceptionally adventurous life of a legendary figure in Italian history. It also covers technical activities, such as those of a water commander who is not very well known, and touches the deepest elements of the story, because the octant was in Garibaldi's possession right up to his death.

The sum of technical knowledge and curiosity about the octant system is connected to the emotional aspect of the exciting life of Garibaldi, who serves as a central figure for students of every Italian school interested in the building of Italy itself.

The second subject that has gained intelligent appreciation is the touchscreen. This is a feature that allows anyone, with nothing but their fingers, to access the innermost soul of the intelligent objects documented and described.

It allows you to simulate movements, or create models, to view valuable drawings over the network.

Touchscreens have become the technology of choice in recent years, because their systems are increasingly inexpensive, they are easy to use and they replicate the flexibility of a mobile, which is, obviously, now a planetary tool used by more than 3 billion people, about 25% of humanity. It follows that the reality of today's knowledge exists through a complete sharing of information and emotional participation in an object, in its physical, historical, emotional and even cultural realities.

On behalf of the Maritime Museum, the Tristan da Cunha plant collects inquiries from potential visitors who show a tender interest in this Atlantic island where a ship full of Camogli sailors sank two hundred years ago, resulting in the founding of a colony that today speaks the Camogli language. Every visitor is struck by the island's history and shares in its emotional reality as they collect information about the museum. Both the logical nature and the emotional sensitivity are satisfied, because the museum is about participation and sharing.

The second application is dedicated to the Museum Collection Paul VI in Concesio, the province of Brescia, which gathers all the works donated to Pope Paul VI by the artists of the world who have visited it. The collection comprises about 4000 works of modern art, and is very beautiful and diverse. The Paul VI Collection is important for its 360-degree form of openness, and the secular nature of each work and gift, despite the deeply religious aura of the surroundings. It inspires actions outside of the spirit of the art collection, such as personal prayers to mystery and religiosity. Each visitor can be immersed in an atmosphere of human culture that is almost metaphysical. Only the DYNAMIC MUSEUMS FOR SOCIAL LEARNING program can offer this pervading presence of the divine and the human through a cell phone, a link and a simple and concise indication, retrieved from the cloud.

The DYNAMIC MUSEUMS FOR SOCIAL LEARNING Project completely changes the path and manner of visiting a Museum, because the works that may please the person are chosen first, along with indication of how best to visit them and spend their time and effort engaging with them, obtaining an experience that is both pleasurable and deep, deliberate and acceptable, and allowing for the building of a true personal culture.

16.4 The Evolution in Participation in Museums

Today, we have at least six billion potential users of Museums, from the elementary and simple museums to the great cultural cathedrals in big cities. These institutions participate in a universal culture, indicate paths visitors may wish to follow, and often respond to changing values. Often, artistic values may correspond to cultural archetypes, certainly acceptable, but not always comprehensive or flexible.

Too often, artists, objects and cultures have eclipsed the time in which they may shock, with values that once only existed in the shade coming into the light worldwide, with maximum transparency and ease.

The reasons for these changes are as follows:

- It allows everyone to observe, learn, and investigate these models, creating possible new insights and innovations, getting to know the objects in detail, infusing their existence with new intelligence that will inform them for centuries.
- It offers the designer of Museums data and models for a reconfiguration of the Museum itself, ever more responsive to the tastes, pleasures and cultural essences of the human visitor.
- It also lets everyone know about the items in storage and archives, which, by law, cannot be made public except with special authorization.
- It enriches the intelligent object's description files, movies, and multimedia for increased knowledge and facilitated redesign.
- It enriches knowledge of the object in relation to the society and the environment of the time.

We have focused our energies on building a futuristic reality whose roots are firmly of today and whose branches represent tomorrow. The force of modern knowledge can overcome all obstacles of time, distance, environment and culture, if well guided by precise objectives. We have accessed all current knowledge on computer design, communication design, data management, virtual and multimedia representations of multivariate mathematics to show how you can draw each person more and more into the reality of the museum. This is an attempt to move away from sole consideration of the paths that are set and programmed, and towards the free choice of knowledge and participation, informed by intelligence and passion. In this way, the visitor finds herself as the new front of the intelligent object, the work of art, the archaeological find, the scientific instrument. The quality of the object draws human attention, in its logic, in its emotional and vital participation. It is a contribution intended to be parallel to the imposed, programmed path, which remains effective as a suggestion for a priori knowledge. Aristotle wrote, in 340 BC: "The fourth kind of quality is made by the figure and the shape belonging to each object; Moreover, it is by its form that each object is said to possess a quality." (Aristotle, categories, 8, 10°) "All the qualities can be called an affective quality, by each of them produces an impression on the senses." 9 ibid, ab.

16.5 The Results

The overall table shows the numerical results of the suggestions that the DYNAMIC MUSEUMS FOR SOCIAL LEARNING Project offers the potential visitor, indicating the set of smart objects that will most likely attract her interest (Figs. 16.2 and 16.3).

The final scoreboard shows the results, so that the potential visitor receives the classification suggested for the linked objects in the project database on her electronic device (computer, mobile tablet or any other system on the network) (Fig. 16.4).

16.6 The Culture of Knowledge and Objects

Even the cultural and emotional parts are filtered through the DYNAMIC MUSEUMS FOR SOCIAL LEARNING Project, as happened with the works of Giotto, which engulfed the Scrovegni Chapel in Padua to work in churches and collections. Who reads the aesthetic works of Croce and Ragghianti, the founders of aesthetic education, recognized today as mental and rigid models that are critical to human culture? How to interpret Haring, who, from the ad hoc galleries of the New York subway, revolutionized visual communication, without rules and dictates? And are today's aesthetic pleasures and the varied tastes they represent, that is, that of every human person, beyond all criticism? And are the works retrieved from

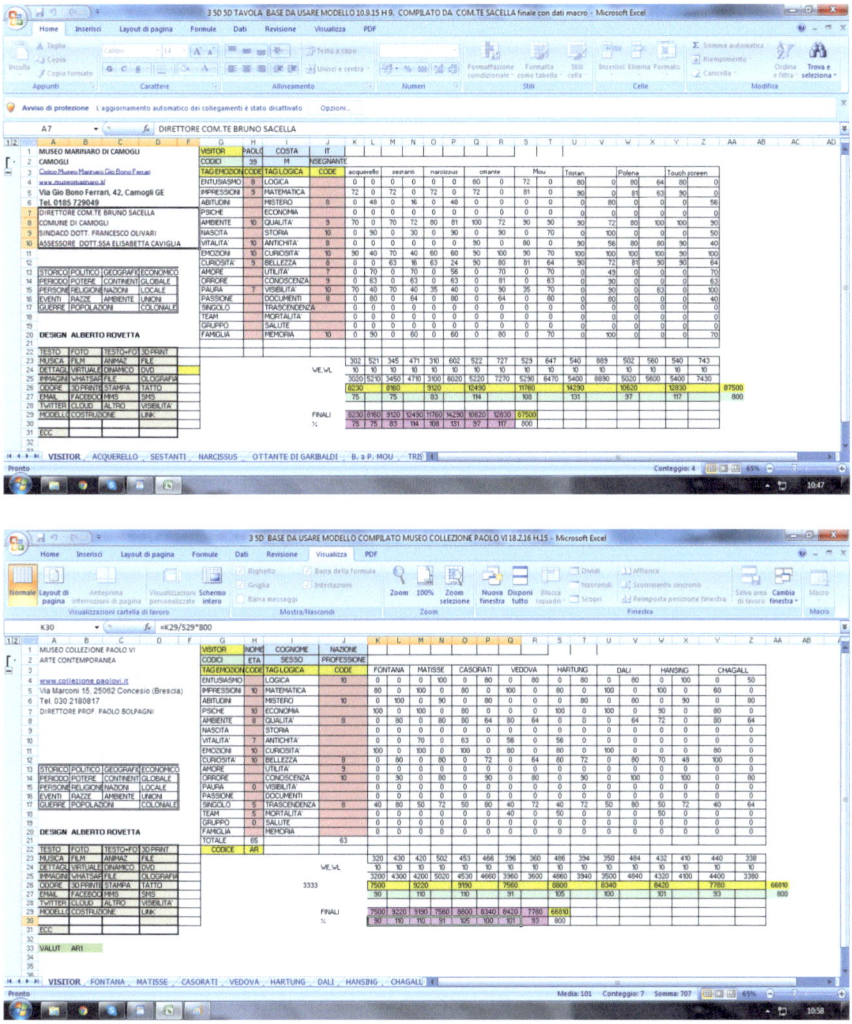

Fig. 16.2 Tables for Scores

FMR magazine, collected by Franco Maria Ricci, incredibly beautiful and often unknown because they are not present in the usual and excavated cultural revolutions, the so-called excellence, which often deeply neglects masterpieces?

We consider therefore that the new Museality, entrusted to the research and tastes of the visitor, is the first step in retrieving a deeper knowledge of our smart objects, the reality of today and yesterday, looking to a shared pleasure in curiosity about beauty. It will also open storage spaces that are often disregarded and forgotten, filled with magnificent works and unforgettable objects, indicating signs of an understanding and universal human participation, and universal means of all

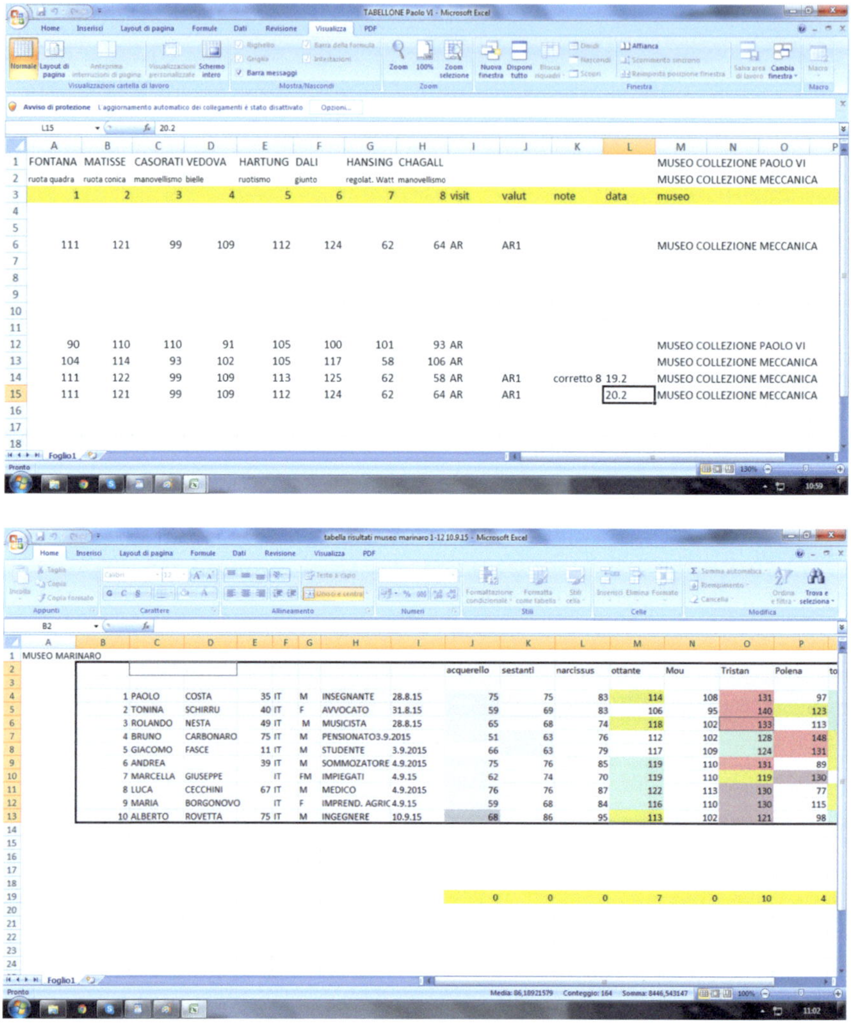

Fig. 16.3 Results of Scores

humanity by race and religion and culture, viewed through the new network. The world of the objects becomes a common heritage for all, without distinction of knowing and being.

The objectives of the "DYNAMIC MUSEUMS FOR SOCIAL LEARNING" program is:

- Exploitation of the "DYNAMIC MUSEUMS FOR SOCIAL LEARNING" Program by the Italian state and non-state museums, for the participatory dissemination of Museum activities and Museum objects in close relation with the cultural and emotional demands of potential visitors, extended to the Italian

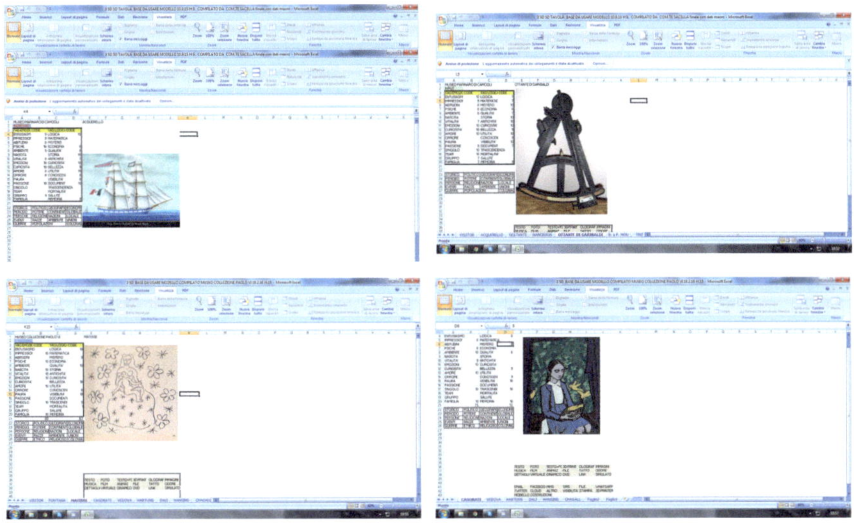

Fig. 16.4 Curator Evaluations

territory, and regarding foreign visitors, for the widest possible enjoyment of the Museums in Italy

- Development of attention to small museums and those not incorporated in full in the Museum tourist routes to strengthen their visibility within the Italian territory, with information and presentation regarding their special cultural interest
- Possibility of collaboration with institutions and foreign companies to promote the use of the DYNAMIC MUSEUMS FOR SOCIAL LEARNING Program for increased visits and cultural tourism in Italy
- Possible participation and presentation of results of the Program to the Expo 2020 in Dubai, with the cooperation of interested institutions, to develop the theme of EXPO 2020: "Connecting Minds, for the Future", and providing continuity with EXPO 2015 in Milan, including regarding the Tree of life.

16.7 A Reference Example

We use the example of the existing, functioning, beautiful, original Exploratorium system, on the beach of the bay of San Francisco. It was built a few years back, enjoying current glory and very high attendance. In the Exploratorium, the exhibited objects express, through movies featuring tables and explanations, their "intelligence," involving the intellect and emotion of the visitor. Figure 16.5 shows some aspects of this. It has destroyed the old-fashioned view of the museum that

Ocean	Connecting minds	For the future	Glasses
Exploratorium, San Francisco (CA), USA, 2015 *Courtesy Collection AR*			

Fig. 16.5 Exploratorium in San Francisco, USA

involves joining the queue to see many objects and considers this line to be the path of truth, without distinction, without criticism, without participation.

Think about what infinite wealth of knowledge could be received by the person who can feel the culture of the selected item, can hear the sounds and see the color graphics and dynamic of the frequency spectrum? It is as if everyday words become music played by a magical orchestra.

16.8 A Possibly Difficult Future

Neither should we forget the loss of national identity that accompanies the millions of people who, in recent years, have transmigrated from their countries of origin towards different continents and countries, in the search for salvation, or a better quality of life. Respect for every human being, especially when in trouble, gives us this wonderful and unexpected image of Museums as the heritage of humanity, its own birth, culture and civilization in which we were born. It remains a testimony accessible from anywhere, especially with the DYNAMIC MUSEUMS FOR SOCIAL LEARNING Program, a mirror of the past and a vision of the future.

Chapter 17
Basic Patent Elements

Modeled on general rules of patents, the idea of patenting the Method to be used to enhance the intelligence of objects (IoT Intelligence of Things, in today's literature) in the DYNAMIC MUSEUMS FOR SOCIAL LEARNING Project.

Patent for DYNAMIC MUSEUMS FOR SOCIAL LEARNING

The basic points of the methods to be patented are:

17.1 Method of Exposure

The METHOD OF EXPOSURE of museum objects should be such that the intrinsic emotional reality and logic that is the prerogative of the human being, and that has been accepted in the latest Museums for the past decade as being a determining factor in the quality of a display, can emerge. The method addresses the psychological, or better yet, neuropsychological, demands of people who want to approach the cultural circuit of museums as sources of historical objects equipped with meaning and significance. When we speak of the "significant literature," this means the material presented that is entrusted with the function of expressing a series of facts, thoughts, stories, events, and feelings, in a defined context. Meaning refers to the sum of the intrinsic and extrinsic information on the Object that perfectly defines its context. The Visitor can retrieve a logical and emotional reality, even one that is shared, through the remote and hidden guidance of the Curators of the Museums. In fact, the Museum Curator (or the knowledgeable person whom they designate for the task) offers intelligent objects infused with significance, as tools through which anyone can become a sharer in new knowledge. The result, seen as creation of a situation in which a person decides to personally and physically visit the Museum, enriches the person and gives a moment of happiness, completeness, a feeling of belonging and growth.

A. Rovetta and E. Rovida, *Scientific Knowledge Communication in Museums*, Research for Development, https://doi.org/10.1007/978-3-319-68330-0_17

The innovation of the patent is that it represents a method of crossing the aspirations/desires/interests/vocations of the Visitor who aspires to view the intelligent objects in the Museum along mature and cultured directions with the foundations of modern neuropsychology that lean on reality and logical emotion. This assessment has been a factor in Museum culture for centuries, and has long been linked to the figure of the educated person who collects intelligent objects, cataloging them, ordering them and proposing an almost obligatory way for them to be experienced. A path/route between objects in a certain predetermined sequence imposes an implicit cultural path that may seem forced and incomplete for the visitor who comes to the museum for the first time. The Method is based on mathematical processes and the gathering of assessments, even those with fuzzy logic and neural network systems, because every experience with the Museum's Items has potential worth; it can be archived, stored and retrieved for new assessments. The result of the method is to present the objects to the Visitor in an innovative way that relates to their modern reality, not as inert things, a collection of corpses from the past, but rather as items that inform the present through their past lives, items that existed for the use and consumption of human beings and for the enrichment of their emotional intelligence, in a participatory way.

17.2 Enucleation Methods

The ENUCLEATION METHOD uses the latest computer technology related to classification, for the distinction of classes of objects, for the analysis of the quality of intelligent objects themselves, for the ability to communicate the historicity of their shapes, sizes and colors, for the clarification of the beauty that attracts us to them. The enucleation method is used to identify those objects that are best perceived, understood, and loved by the visitors, according to the preferential indications that they express in a digital table. The Guests can thus tailor their visits to their own characters, intelligence, sensitivity, requesting that the Italian museums (and later those in the rest of the world) show them only those objects that are likely to please and interest them most. But the enucleation needs a crossover method between different families of interest, because whoever exhibits the objects and seeks to give them a meaningful and participatory meaning must bring forth and flesh out the specific properties of every intelligent object. The significant properties go beyond the mere history, whatever it may be, relying heavily on the nature expressed by the meaning that the object represents, a meaning that, if emphasized, will become loved, and then further expressed, by the visitor. The enucleation method is conducted by determining the numerical values for preference and participation assigned by the Visitor to each intelligent subject, setting a ranking for each museum and making allowance for the specific preferences of each person for each Museum. In this way, you get an extremely wide numerical ranking of items enucleated, in an orderly fashion, which are arranged in rank according to the positions occupied in the possible preferences and qualifications proposed by the Curator of each museum. The enucleation is sent to the computer through the network, and is then passed on to the digital systems of every possible Visitor, who can then see every possible subject, rediscover the features that were less distinct in the first digital chart, and perform their final classification. Particularly important is the fact that every object can be provided as information to the possible Visitor, before or after the visit to the Museum, in the form of printed paper, a figure, a file, an object printed in 3D by a 3D printer, or a hologram.

17.3 Method of Sharing

The integrative method is made possible by providing the visitor with all the cultural components, logical, emotional, passionate, that a smart object can express. This method makes the Smart Object an active part of the participatory visit to the Museum. The result is the choice made by the DYNAMIC MUSEUMS FOR SOCIAL LEARNING Program's computer control as to the most significant Object, with a numeric score which has the task of creating a ranking. The subject is selected later.

The method relies, in a large capacity, on the cultural and emotional experience of the Museum Curator. He/she should be an expert in the history of the objects, but above all capable of experiencing the cognitive and emotional bond that stems from the viewing of Exhibits, so that the emotion and understanding of beauty and substantial wealth of the Object can be passed on to the visitor. This method is a clear revolution in today's Museum culture, in which there are very few (almost null) references to this Museum setting.

17.4 The Goal: A Method Suitable for Enabling Complete Participation

The subject is the logical and emotional counterpart of the Visitor, who may have very intense cultural and cognitive expectations that express a logical and emotional choice relative to said expectations. On a phone (or similar device), you can display a 3D image of the object with icons and no writing, simple, comprehensive, with lights and contexts that envelope the image, in a way adapted to the probable Visitor. It is the goal of a complete vision, which is adaptable to and owned by the visitors, who choose the colors, backgrounds, perspectives (a task well effected by today's computers) in a way that adapts the subject to their own preferences and in the context of exhibition that speaks to their own human reality, in its logical and emotional aspects.

The method provides the probable Visitor with an Object that becomes an acceptable figure to be shared, through interpretation of its main aspects. The phase of understanding and knowledge, and even love, comes later. The method is a very welcome one, because it is a prelude to the visit, without requiring the visitor to become emotionally engaged before the visit, only providing them with anticipation of the emotional and physical environment in which the subject is immersed.

17.5 Methods of Emotional Discrimination

The EMOTIONAL DISCRIMINATION method is a training tool for Museum Curators that enables them to complete their cultural background through new developments in neuropsychology that offer better understanding of human psychic action, as created through cooperation of the two cerebral hemispheres. Modern science thus returns to the importance of emotion and logic, which, in their union, represent the many facets of human personality and behavior. Because each Museum Curator may be inclined towards different emotional evaluations, the emotional discrimination method adopts a kind of emotional TRAINING program: packages of objects are presented, with indications as to the EMOTIONAL values reported by psychiatric experts. In the first set of 8 objects, the emotional values and the determined logical values that serve as the baseline of the DYNAMIC MUSEUMS FOR SOCIAL LEARNING Program are shown. Having completed the first cycle of 8 objects, 32 objects appear in succession, merely displaying their form, without any indication of their emotional and logical value. The Curator-in-Training, the precise beneficiary of the training method described here, chooses values, ranging between 0 and 10, and immediately compares them with the system's standard values that the control computer has displayed alongside their own. The method is progressive and teaches the prospective Curator how to work on the proposed objects; it is suggested that it be applied until the Curator no longer reaches a percentage of positive results equal to or greater than 75%.

17.6 Methods of Reconfiguration of the Museum

This method consists in the RECONFIGURATION OF THE MUSEUM, specifically of 8 objects, inserted into each table programmatically, that have been precisely catalogued by the DYNAMIC MUSEUMS FOR SOCIAL LEARNING Program in the first table. The data collected from the Visitors who have chosen the objects are saved in a Program file marked PREMIUM, which, for each subject, indicates (A) the relative score and (B) the position on the chart for each Visitor. The method counts access and the average position in the rankings for all guests.

In the end, for each museum, you have a ranking of the favorite items. From the ranking—selected from first to last—it can be deduced which objects must be presented and in what context. The context must be built/rebuilt in a simple and sufficient way, to describe the actual reality in which the subject has been expressed, or expresses itself, as efficiently as possible.

For the method, it is useful to have a concrete idea of the numerical value of the exhibited objects, based on how many Visitors focused their visit on those items. It is necessary to know the favorite of as many possible Visitors, but users of the DYNAMIC MUSEUMS FOR SOCIAL LEARNING Program.

The program becomes a tool for selection shared by experts and validated by the success it achieves among Visitors, who may express their comments in a simple method (email, WhatsApp, Facebook, etc.).

17.7 Acceptance of Loving Method

This method intends to fully verify the validity and effectiveness of the proposed selection, referring to the limiting conditions, as is proper in any heuristic method applied to such a complex notion concerning the human being as neuropsychology. The method involves envisaging an ideal theoretical element, that is, a VIRTUAL VISITOR, who presents a single type of emotional value, with the maximum numeric value being 10. The DYNAMIC MUSEUMS FOR SOCIAL LEARNING program runs the data from all the Museums through the database, allowing them to read the answer of every museum, noting which are the favored subjects and which preferability indexes have been assigned by the Curator of the Museum. At this point, you can check with a test, also conducted—if you will—remotely, as to whether the selected objects have a reasonable match with the values shown by the test. You will need the advice of three experts, according to an Expertise Evaluation Team. The results are processed and will be maintained, as wealth in the database and as sensitive data in the working proposal of the DYNAMIC MUSEUMS FOR SOCIAL LEARNING Control Center.

17.8 Fast Convergence Method

This method is proposed for the very fast data management of museums that are in the DYNAMIC MUSEUMS FOR SOCIAL LEARNING database. In this way, there is a complete, albeit crude, set of favorite items chosen by users. The method crosses the Visitor's data with the data from each museum, in the tables of 8 different objects, selecting only the top 3 results and ignoring the others. The final printout is rough and yet contains the users' preferred results. These data will later be crossed with the preferential data from the Visitor on the size/location of the museum (whether it is big/small/single-subject/on larger premises/isolated/etc.).

You can then re-enter the DYNAMIC MUSEUMS FOR SOCIAL LEARNING Program, which is complete after having obtained a reasonable and positive introductory message on the choices made, according to the demands of the visitor, who is always the deep and real protagonist of any visit to a Museum, for the betterment of life.

Chapter 18
Hemispheric Asymmetries in the Brain

Brain activities regarding knowledge and behavior are reflexive of scientific steps and human knowledge. Behavior is explored deeply only after scientific measurements of brain actions in both hemispheres.

In these pages was born one of the most intense analyses on all the ways that the two hemispheres of the brain affect human behavior.

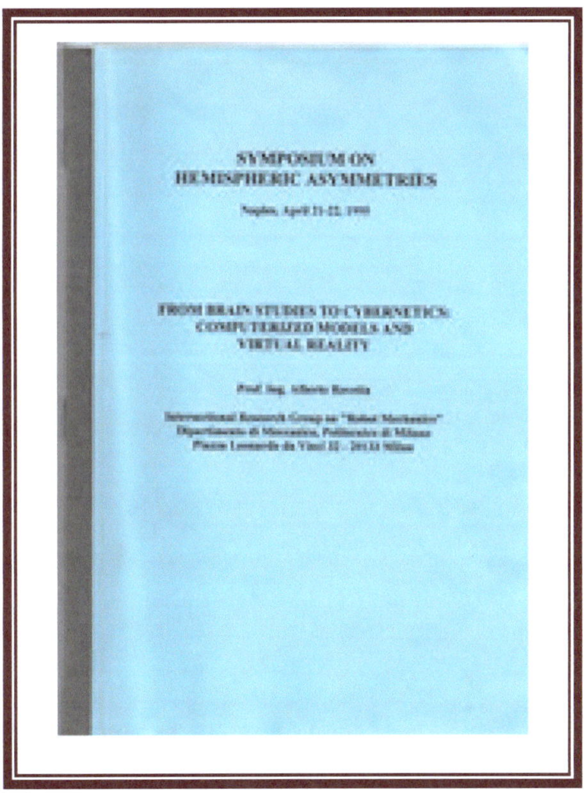

A. Rovetta and E. Rovida, *Scientific Knowledge Communication in Museums*,
Research for Development, https://doi.org/10.1007/978-3-319-68330-0_18

Part VII
Project "5 Museums"

Chapter 19
Project "5 MUSEUM" Presentation

The recent neurobiology has learned the different potentials of the human brain in the left hemisphere, the logical part, and the right hemisphere, for the emotional part, expressed by human behavior. The Project DYNAMIC MUSEUMS, first, uses these principles to make happier and complete the visit to the Museums, offices of knowledge and today's memory.

19.1 Results of the Tests of 5 Museums

The Project applies to 5 National Archaeological Museums, indicated by the Ministry Mibact: ALTINO, ANCONA, EGNAZIA, FERRARA, SIBARI. It puts the visitor at the center of the visit, as it crosses the personal data of the Visitor's behavior with the basic information that the Museum Curator assigns to objects. By use of new knowledge of neurobiology and of today's artificial intelligence with "deep learning" ("deep learning" in the language of today's artificial intelligence).

The Project DYNAMIC MUSEUMS born at the Politecnico di Milano, sponsored by the Fondazione Politecnico di Milano, and is developing a convention with the Ministry of Cultural Heritage and Activities and Tourism, Mibact, to apply the project to 5 Italian National Archaeological Museums.

Testing have been conducted in 5 National Archaeological Museums, with the use of a computer platform, and acts in a tablet, given to the museum. The program asks the visitor to describe his state of mind through two figures in "diamond." In them, the visitor expresses his mood, giving the screen a numerical value (between zero and ten) to 8 logical and 8 emotional parameters. The program crosses the data of the Visitor with the indices assigned by the Museum Curator to 16 selected objects from the Museum. The program represents the objects corresponding to the wishes of the visitor.

A. Rovetta and E. Rovida, *Scientific Knowledge Communication in Museums*, Research for Development, https://doi.org/10.1007/978-3-319-68330-0_19

The results are very important because they concern 1021, i.e. more than 1000 applications, the Project in the National Archaeological Museum 5, indicated by Mibact, from 2 August 2016 until the day September 3, 2016.

The Project DYNAMIC MUSEUMS works with an IT platform, which can be modified by the Scientific Center of Project DYNAMIC MUSEUMS. The screen shows two “diamonds” to 8 edges; the left represents the action of the left hemisphere (logic), the right is the action of the right hemisphere (emotions).

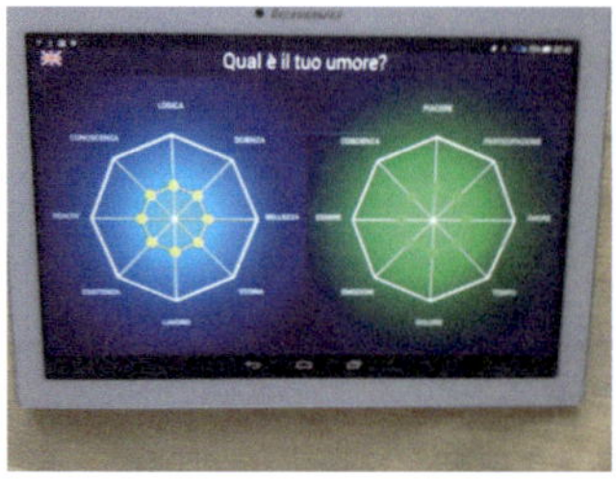

This image appears when the program starts and the Visitor, with the tip of a finger, moves gently and easily each of the circles of the 8 parameters of each “diamond”, stopping on the desired value (minimum 0 at the base, up to 10 outside). The choice of the meanings and values depends on the emotional and logical moment of the Visitor. The “diamond” Left is the 8 values that define the logical action of the left hemisphere. The beauty is part of the logical structure because it can also be understood as a “model” of thought. The “diamond” Right represents the emotional significance of perceptions and sensations. Remember that these schemes are valid for each museum, intended as a seat of knowledge and memory. The Museum is the site that the visitor wants to access in a pleasant and welcome way, waiting for emotions and pleasure. The emotional part is very important in the years around 2010, because today the world lived always expresses more emotion, expectations, hopes, efforts. The ‘Diamond’ emotional has never been used before, even in the cognitive process, and it is considered a feature of priorities of Project DYNAMIC MUSEUMS.

The “diamond” receives data from the Visitor, and crosses them with data evaluation data from Curator. It is a remarkable work of software, algorithms and weights in a multivariate mathematical. The outcome of claims as a guest on his state of mind appears as magnitudes that the visitor has chosen to indicate the values he has chosen, from 0 to 10 (in fact some were left at 0). You notice the sign to the two hemispheres of the brain, which are the basis of his behavior. (The data and results are covered by complete privacy). Here starts the intersection of Visitor Data with the reference values indicated by the Museum Curator for each object. There is the initial table that the visitor can fill out in detail or simply with “I do not answer.” It includes age, sex, education, work. It can be developed differently for each targeted application and the “modern artificial intelligence models” will guide the visitor.

As requested by the Ministry MiBACT, you want to know the "confidence" of the Visitor with the Museums in specific visit.

The program developed with the Convention with the Ministry of Business, Cultural Heritage and Tourism has affected 5 National Archaeological Museums: Altino, Ancona, Egnazia, Ferrara, Sibari. The indications of the Visitor and numerical notes of each Museum Curator create the indication of three "smart objects", first among the 16 selected figures from each museum. At this stage, you are only using 16 items, which can be multiplied very easily via software and via database.

Results of the tests of 5 Museums:

- The average approval rating was 8.24 points out of 10 (scale of customer satisfaction);
- 56.5% reported preferences for the logical characteristics;
- 43.5% reported preferences for the emotional characteristics;
- The percentage of visitors up to 35 years was found to be 33.4%;
- The percentage of visitors aged 35 to 55 and over was found to be 58%;
- Gender was represented by males at 49.7%, and females at 42.6%, with 7.7% declining to answer;
- The declared professions are as follows: Employees 26%; Students 19%; Teachers 14%; Freelancers 14%; Other 27%;
- The frequency of visits to museums, in general, appears to be less than 5 times a year in 61% of cases;
- The frequency of first-time visits to the above-mentioned 5 Museums is found to be equal to 75%.

The program is launched with a click, and the design platform returns a ranking of 'smart objects' falling within the field of Visitor Popularity, with a numerical display of the evaluation in the Visitor's optics.

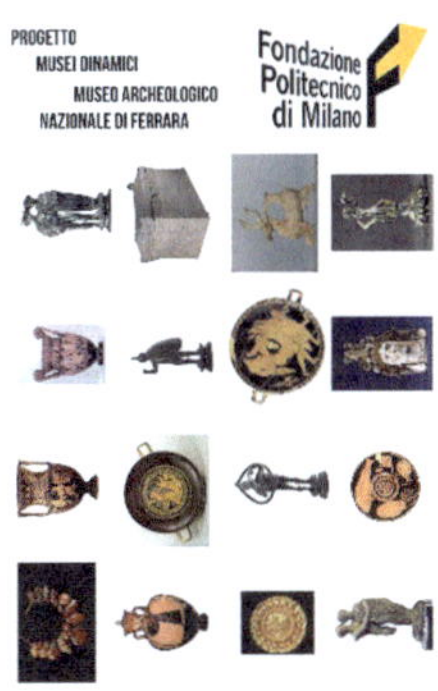

This figure represents 16 objects specified by the National Archaeological Museum of Ferrara.

This example concerns an application related to emerging smart objects, with indication of the numerical value reported by the Visitor. Every visitor usually gets his own personnel ranking, according to the logical and emotion data that regulates his behavior.

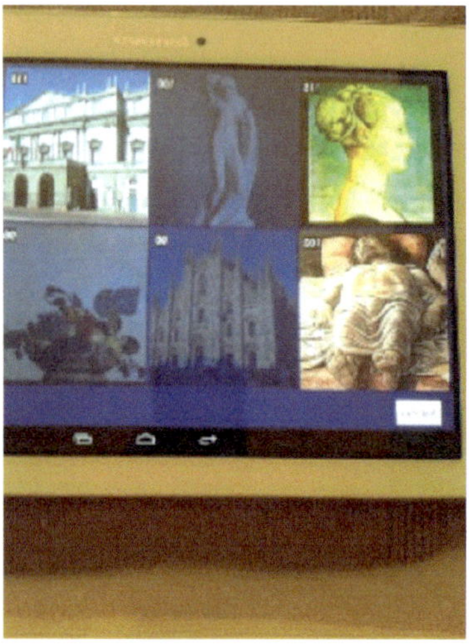

This result shows the figures suggested for a visit to a city (in this case, Milan), with links to specified objects and possible indications on the means to view them.

After the presentation, it follows the basic display on the principles of the DYNAMIC MUSEUMS Project.

The DYNAMIC MUSEUMS Project is integrated into cultural development based on new technological and cognitive methods, with recent models of learning and maintaining knowledge.

- The DYNAMIC MUSEUMS Project has been developed under the auspices of the MILAN POLYTECHNIC. FOUNDATION
- The DYNAMIC MUSEUMS Project offers new benefits for the social and cultural aspects of the use of museums in Italy
- Social and economic aspects are at the forefront of development of the application of the DYNAMIC MUSEUMS Project
- The great, rich cultural, historical, artistic and social aspects of Italian Museums can be revised and designed through new methods of use: The DYNAMIC MUSEUMS Project excels in making the riches of the Italian Museum system more exploitable.

19.2 "Dynamic Museums" Project: Sponsors and Initial Support

- Knowledge management touches all people, Italian or otherwise, who are affected by the beauty of the works and objects that are present in Italian museums.
- The development is brought about by Italian and foreign participation in the creation of a new reality in the Italian Museum system.
- The Project's Sponsor represents the economic and financial realities, linked to the policy of cultural knowledge of the treasures in Italian museums.
- Sponsors can give financial support to the launch of the project, which, using applications and attendance systems, will support itself after start-up.

19.3 "Dynamic Museums" Project: Application to the Italian System

- Implementation of the project consists in getting the application onto any phone or programmed electronic system that will allow the visitor to prefigure and build a potential personal program of visits to Italian museums. It uses a digital platform that can be installed on any digital system.
- The models used correspond to the visitor's interests and individual personal tastes.

TABLE OBTAINED BY THE POTENTIAL VISITOR

MUSEUM CURATOR TABLE WITH ASSESSMENTS

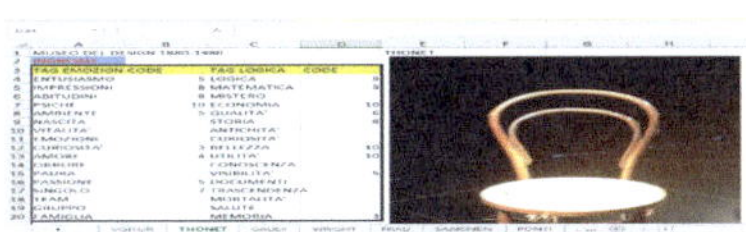

APPLICATION OF ALGORITHMS FOR MODEL OF ENJOYMENT AND ASSESSMENT

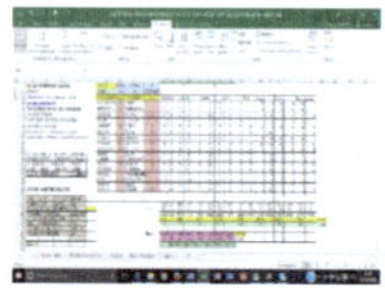

19.4 Methods of the "Dynamic Museums" Project

Through mathematical algorithms, the DYNAMIC MUSEUMS Project generates a numerical-physical platform of suggested results to the POTENTIAL VISITOR; out of the ranking, the most pleasing and interesting museum objects that match the visitor's intent are then listed.

The result for each visitor can also be reproduced as a souvenir, through printing, photos, holography, 3D printing, getting the proper documentation of those museum objects that more like and interest to the visitor, which can be collected at the end of the visit.

PRODUCT EXAMPLE WITH 3D PRINTER FOR THE MUSEUM VISITOR

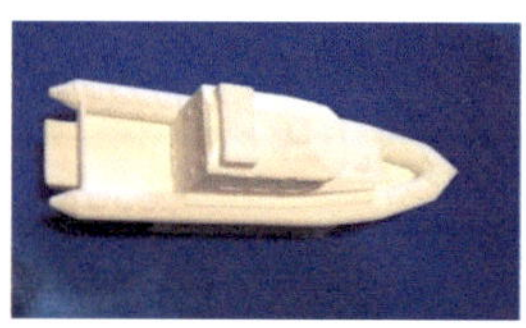

- The DYNAMIC MUSEUMS project radically transforms the use of Museums involving the POTENTIAL VISITOR by engaging his personality through its logical cognitive aspects and through his emotions, expressed in daily life by the actions, respectively, of the left and right sides of the brain.
- You can use computers, cell phones, tablets, or any other type of communication device, as well as a computer station located in the Museum itself.
- The DYNAMIC MUSEUMS project enhances the convergence of logical/ emotional points of interest. The platform is also used for the reconfiguration of museums themselves or that of museum exhibitions, which can be organized for the various categories of Visitors, as well as prepared for every individual visitor.
- THE NEW SERVICE OFFERED by the DYNAMIC MUSEUMS Project for the new use of Museums takes place through identification of paths between multiple museums, offering all the information online necessary to reach and

access these institutions, as well as that of all the nearby facilities that will make for a more convenient, comfortable and appealing visit.
- In international development, the project will take on the name of "EASY VISIT," as a way of expressing the simplicity and the pleasure to be found in the new mode of museum visit.

19.5 New Professionals

The professionals that are needed are:

- Recent college and high school graduates educated in the disciplines related to cultural heritage, to reinforce the links between museums and the culture of citizens with a new urgency
- Graduates well-versed in computer science, to manage the data and documentation about the Museum's objects simultaneously (including regarding smaller museums, redesigning and reconfiguring the museums themselves, based on the experiences collected from visitors)
- Recent college and high school graduates well-versed in economics and the management sciences to stay on top of such political issues as budget cuts and economic recovery and innovation in the management of museums, and to play a positive supporting role in decisions regarding how visitors will be guided through the museums, while today it is pure surveillance (where a micro camera could do the same job, more accurately)
- Recent college and high school graduates well-versed in tourism sciences, to rebuild the Italian tourism map and culture, not leaving out any existing museums
- Curators of Museums, who, through new methods of information technology and communication, can update the culture in a valuable way
- New graduates well-versed in the use of holography and 3D printers, to make sure the product is of high quality
- Recent college and high school graduates well-versed in foreign languages to promote the internationalization of small Italian Museums.

19.6 New Enjoyment of Lessons

It is necessary to propose an integration in the use of lessons for the following disciplines:

- Interior Design, for definition of the routes in the museum, the internal distribution of objects, the design of the furniture, etc.;
- Communication Designer, for creating graphical, interactive media, the internal signals, the publications that support the operation;

- Designer specializing in Design Services,
- Technical experts in the digitization of paper documents,
- Technical experts in three-dimensional digitization,
- Technical experts in digital techniques, or traditional techniques, for construction of three-dimensional objects.
- In addition, based on the above-listed professionalism, it is reasonable that those with a Degree in Museology would be preferred.

19.7 Business Plan

1. Research and Development
   a. Development of working prototype
   b. Application to more Museums
   c. Drafting protocols and instructions
   d. Function test
   e. Demo and local and international presentations
2. Creation of the Product
   a. Design with appeal
   b. Using symbolic icons
   c. Application with outputs printed in 3D, holograms, etc.
   d. Evaluation of the application and operating costs
3. Development of the general organization and recruitment plan
   a. Cooperation with IT firms for online assistance, and in every application
   b. Promotion to foreign users and application in all the primary languages
   c. Promotion to large organizations with the support of both Italian and foreign institutions
4. Protection of intellectual property and patenting
   a. Drafting of the patent on the cooperative method applied in Italy and preparation of the European patent
   b. Drafting of the patent presentation in targeted and restricted areas for possible international development towards the Italian Museums
   c. Any new patenting innovative after results
5. Marketing, implementation, promotion
   a. Comparison with products with similar intent on the national and international markets
   b. Evaluation of the implementation of the DYNAMIC MUSEUMS product with new possibilities and greater openness

   c. Application not only to museums, but also to shows, theaters, Sporting Events
   d. Promotion to youth and even small children
   e. Promotion in the world at both a high intellectual level and the basic education level through creation of circles and participatory structures at a high emotional level
   f. Overseas promotion of the Reality of the museum and of Italian life

6. Fundraising and sales

   a. International co-operation with sponsors to create an international model for the Museums and the cultural reality of Italian entertainment
   b. Sales in the most widely effective ways.

19.8 Patent Deposit

The "DYNAMIC MUSEUMS" Project it is under patent filing.

Chapter 20
"5 Museums" Project: National Archeological Museums of Italy—MiBACT Ministry—Museums of Altino, Ancona, Egnazia, Ferrara, Sibari

A. Rovetta and E. Rovida, *Scientific Knowledge Communication in Museums*,
Research for Development, https://doi.org/10.1007/978-3-319-68330-0_20

20.1 16 Objects for Each Museum

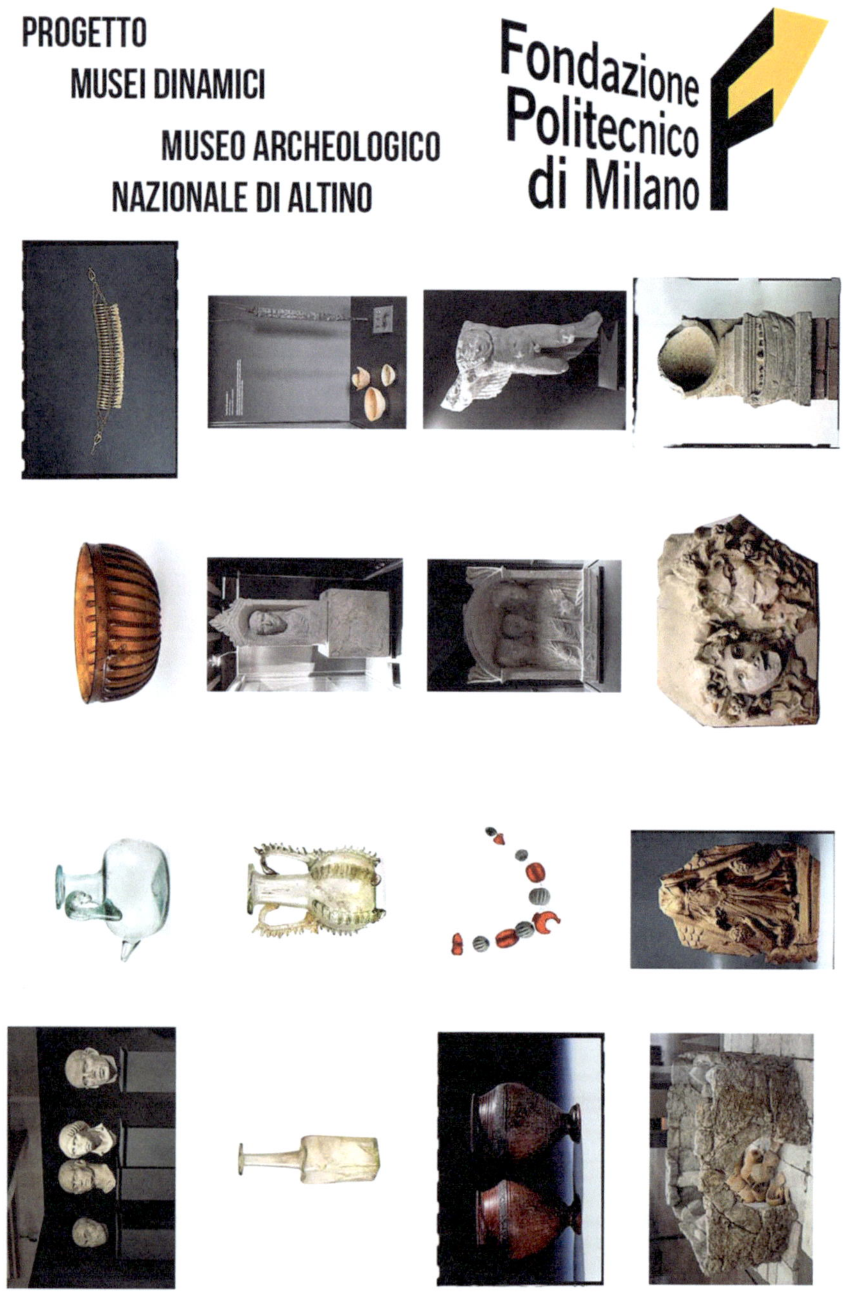

PROGETTO
MUSEI DINAMICI
MUSEO ARCHEOLOGICO
NAZIONALE DI FERRARA

Fondazione
Politecnico
di Milano

PROGETTO
MUSEI DINAMICI
MUSEO ARCHEOLOGICO
NAZIONALE DELLE MARCHE - ANCONA

Fondazione
Politecnico
di Milano

PROGETTO
MUSEI DINAMICI
MUSEO ARCHEOLOGICO
NAZIONALE DELLA SIBARITIDE

Fondazione Politecnico di Milano

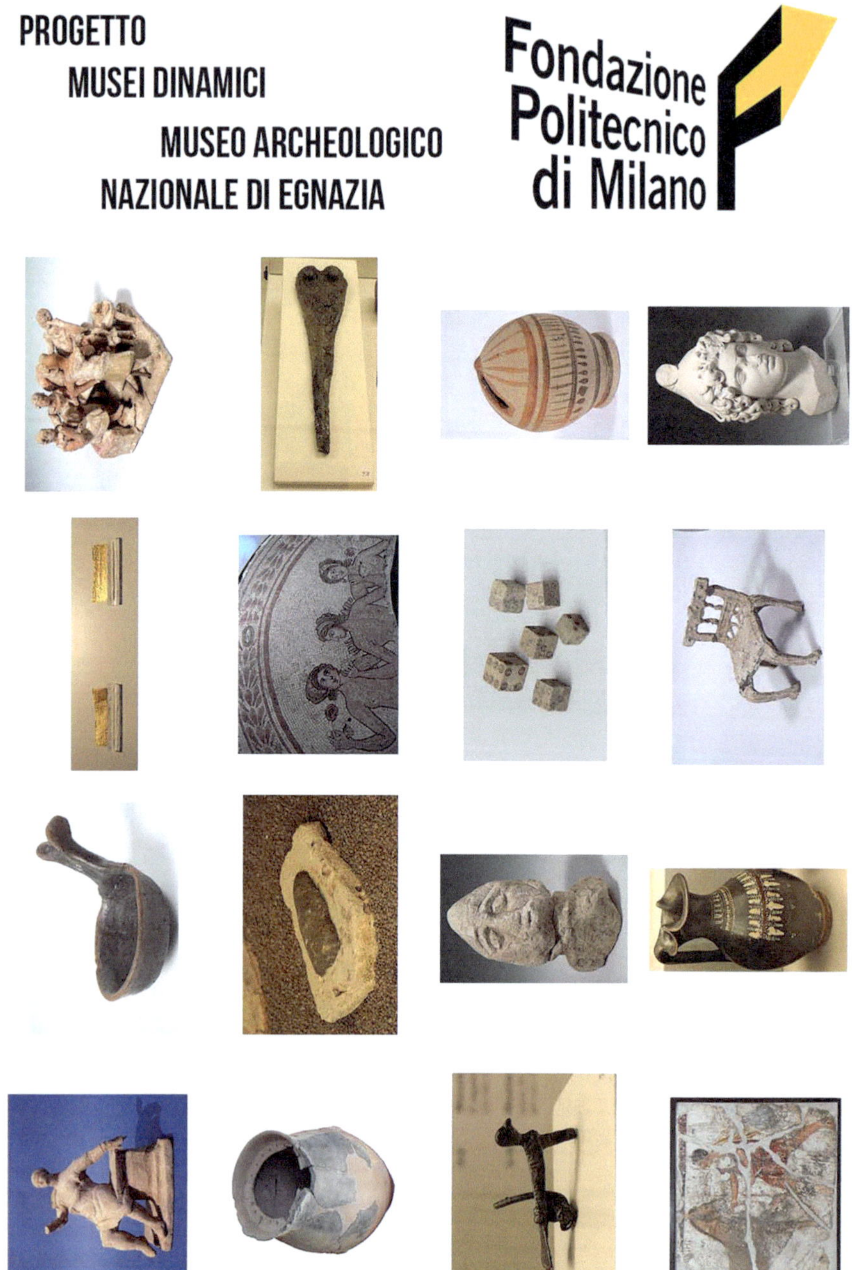
PROGETTO
MUSEI DINAMICI
MUSEO ARCHEOLOGICO
NAZIONALE DI EGNAZIA
Fondazione
Politecnico
di Milano

20.2 Development, and Results Obtained

The data in this section refer to the period of the official trial development, called the "canonic."

1. Results

The number of visitors divided by Museum is expressed in Table 20.1.

The assessment given by the visitors is broken down in Table 20.2 indicating a remarkable approval rating.

The votes in favor of logical characteristics are given in Table 20.3.

THE DYNAMIC MUSEUMS project was applied to five archaeological museums chosen by the Ministry one each in Altino, Ferrara, Ancona, Egnazia, and Sibari. A method of testing was developed to provide the clearest and most positive results, even for subsequent developments .

The project turned out to be a particularly good fit with Museums, helped in great part by the fact that it was carried out with the great support of the museum curators. We found a great partnership in every museum, great patience toward the

Table 20.1 Number of visitors

Museum	Number of visitors	%
Altino	138	13.5
Ancona	132	12.9
Egnazia	179	17.5
Ferrara	45	4.4
Sibari	527	51.6
Total	1021	

Table 20.2 Average scores

Museum	Average score
Altino	8.19
Ferrara	0
Ancona	8.90
Egnazia	7.14
Sibari	8.74

Table 20.3 Votes for logical characteristics

Museum	Votes for logical characteristics
Altino	6.551 (24.6%)
Ferrara	1263 (4.75%)
Ancona	4.140 (15.5%)
Egnazia	7.533 (28.3%)
Sibari	7.122 (26.8%)
Total	26.609 (56.5%)

management of results and preparation of the information that would allow us to bring a complete picture to the Commission.

We discovered the deep knowledge of objects that exists within these institutions, the great love of their museum curators, the will to share and communicate with the people who can then become the protagonists of their own museum visit. The values of appreciation and valuation turned out to be very positive, even above 80%, even though the project is completely new, still unfamiliar, and in a certain sense, still not programmable. The realized platform is equipped with software that is elementary, fundamental and powerful. Software sheets are being prepared that will make for a formidable evaluation platform to further support the Visitor. We want to offer Italian culture in a way that is simple and accessible, and we want to do so for one, one thousand, ten thousand museums, so that the reality of Italian cultural is available for global participation by anyone. In every part of the world, you will be able to download the model from the platform, so that you can visit museums according to your desires, your mood, your own need for culture and participation. The report also lists all numerical data. The opinions of the editors and the people who have followed the project are attached, along with the views of the more than 1000 visitors who contributed assessments. An examination of these opinions clearly demonstrates that visitors feel gratitude for the computer support that provides them with advices and suggestions that allow them to fully enjoy their visit.

Related to emotionality, the personal results revealed a significant new aspect of the relationship between the smart things and the visitor to create a feeling of happiness and well-being in everyday life.

The report is intended to highlight the cultural richness of the Italian reality, in the specific case of the five national archaeological museums, creating inspiration for life, feelings, knowledge, and a sense of happiness for each person who participates in it. It appears to be an opportunity for the cultivation of simple, human well-being, full of emotional growth. The hope is that the project can be expanded to 50 other national museums, enriching the platform software in the process as well. You can download applications, according to the instructions that will be given by the Ministry to the 50 museums. The new trend of artificial intelligence can serve as an important basis for knowledge in areas of national development.

We thank all the people, curators, and visitors who have made a huge contribution to the final project results, which are shown here.

Fondazione Politecnico di Milano

List of Archaeological Museums included in the National Convention Mibact and Politecnico di Milano Foundation for the DYNAMIC MUSEUMS FOR SOCIAL LEARNING Project.

1. National Archaeological Museum of Altino—Dir. Dr. Mariolina Gamba
2. National Archaeological Museum of Ancona—Dir. Dr. Nicoletta Frapiccini
3. National Archaeological Museum of Egnazia—Dir. Dr. Angela Ciancio
4. National Archaeological Museum of Ferrara—Dir. Dr. Paola Desantis
5. National Archaeological Museum of Sibari—Dir. Dr. Angela Tecce.

20.3 Advantages of the Dynamic Museums Project

Following the ministerial rules, we list the primary and most important advantages that emerged from the experimental tests carried out since August 2, 2016 below.

1. Benefits for Visitors
   The visitors, as hoped, have shown appreciation for some of the primary advantages of the DYNAMIC MUSEUMS project, first and foremost of which is the personal satisfaction of being put at the center of the museum, serving as the protagonist of the visit. They were especially appreciative of the proposed tour routes, fashioned according to the behavior and character of the Visitor, and not simply set out as a predetermined sequence of objects. The Visitors' appreciation for the DYNAMIC MUSEUMS Project recognized it as a valuable tool for the optimization of the visit, with a chance to see what is interesting in accordance with the personal knowledge and emotion of the individual.
2. Benefits for Officers of Museums
   The Curators of museums have also expressed their appreciation for the program, eagerly promoting the application of the proposed model on a simple tablet and touting the advantage of having been able to identify the most "popular items," which, when it comes time for the Museum's reconfiguration, can be addressed with greater emphasis.
   A second, very useful advantage obtained from the Project is the way in which it has helped confirm to the Museum Curators the importance of the emotional aspect, in addition to the logical and the cognitive, to make the visit beautiful and enjoyable.
3. Benefits for the Country and for Tourism
   The network application offers the important benefit for tourism and for the territory in general of an expanded presence. It can also attract new fans interested in visiting the territory, which can in turn lead to the development of new feelings of loyalty and a sense of belonging. Beyond that, we must consider the new jobs that will be created as part of the Cultural Heritage aspect of the extensive application of the project.
4. Development for the DYNAMIC MUSEUMS Team. The quality of the project may encourage its application within various National museums.

20.4 Dynamic Museums Project (Convention MiBACT/ Fondazione Politecnico di Milano)

20.4.1 *Premise*

The Centro Cultural Heritage Politecnico di Milano declares:

- Cultural Heritage is a strategic resource for both social and economic development.
- Enhancement of this heritage requires the involvement of a broad spectrum of disciplines and technologies.
- Among the main thematic areas that require interdisciplinary collaborations, we can include: (A) Management and development of models and methods for innovative and sustainable value creation and the promotion of heritage. (B) Communication and information systems capable of spreading and transferring the results of technological innovation for the benefit of cultural heritage, in general, through multimedia products. The foregoing will enrich the knowledge, dissemination and preservation of the heritage.

Thanks to the patronage of the Foundation, it has now been shown that the DYNAMIC MUSEUMS Project (see the results of the 5 National Archaeological Museums selected by MIBACT) is the most appropriate vehicle for achieving the above objectives. The activation of the "game," a.k.a. the "Easy Visit Game," by tapping into the behavioral and mental maps of the people makes the latter the protagonists of an extraordinary visitor experience.

20.4.2 *Results from Tests in 5 Museums*

1. The average approval rating of the E.V.G. was 8.24 points out of 10 (scale of customer satisfaction);
2. Preference for the logical characteristics were reported by 56.5%;
3. Preference for the emotional characteristics were reported by 43.5%;
4. Visitors up to 35 years of age represented 33.4% of respondents;
5. Visitors aged 35–55 and over represented 58% of respondents;
   Gender of respondents was calculated as 49.7% male and 42.6% female, with 7.7% declining to answer;
6. The declared professions of respondents were as follows: Employees, 26%; Students, 19%; Teachers, 14%; Freelancers, 14%; Other, 27%;
7. The frequency of museum visits, in general, appeared to be less than 5 times a year in 61% of cases;
8. The percentage of first-time visits to the above-mentioned 5 Museums was found to be 75%.

20.4.3 *Final Conclusions*

The project, thus tested, shows that much can be done regarding such goals as enhancing the heritage, increasing the number of visitors (estimated + 4% per year), improving and expanding the selection of the most significant objects, and certainly fostering a positive impact on the inflow of revenue (tourism) and economic growth (new hires) of the area that utilizes it. The challenge is to establish the extent and distribution of resources to bring the project on at full speed while maintaining the constant basis of the strategic guidelines already outlined.

20.5 The DYNAMIC MUSEUMS Project up to September 3, 2016

General observations

1. Introduction

The number of visitors is expressed in Table 20.1.

This indicates a very high number of visitors, albeit with significant differences between the museums.

The assessment given by the visitors is broken down in Table 20.2, indicating a remarkable approval rating.

Table 20.3 shows the total number of votes in favor of logical characteristics.

Table 20.4 shows the total number of votes in favor of emotional characteristics.

Regarding the profile of visitors, age distribution is shown in Table 20.5

Gender is expressed in Table 20.6

The title of the study is in Table 20.7.

The distribution of occupations is given in Table 20.8.

The frequency of visits to museums is in Table 20.9.

The frequency of visits to these specific museums is expressed in Table 20.10.

Table 20.4 Votes in favor of emotional characteristics

Museum	Votes in favor of emotional characteristics
Altino	4.505 (22.1%)
Ferrara	1.130 (5.5%)
Ancona	3.317 (16.2%)
Egnazia	5.907 (29.0%)
Sibari	5.557 (27.2%)
Total	20.416

Table 20.5 The distribution of visitors by age

Museum	Up to 18	19–35	36–55	Over 55	No answer	Total
Altino	12 (8.7%)	10 (7.3%)	55 (39.9%)	44 (31.8%)	17 (12.3%)	138
Ferrara	3 (6.7%)	9 (20.0%)	11 (24.4%)	9 (20.0%)	13 (28.9%)	45
Ancona	3 (2.3%)	37 (28.0%)	37 (28.0%)	27 (20.5%)	28 (21.2%)	132
Egnazia	12 (6.7%)	42 (23.5%)	80 (44.6%)	44 (24.6%)	1 (0.6%)	179
Sibari	84 (16.0%)	128 (24.3%)	188 (35.6%)	101 (19.2%)	26 (4.9%)	527
Total	114 (11.2%)	226 (22.1%)	371 (36.4%)	225 (22.0%)	85 (8.3%)	1021

Table 20.6 The distribution of visitors by gender

Museum	Male	Female	No answer	Total
Altino	63 (45.7%)	62 (44.9%)	13 (9.4%)	138
Ferrara	16 (35.6%)	16 (35.6)	13 (28.8%)	45
Ancona	52 (39.4%)	52 (39.4)	28 (21.2%)	132
Egnazia	95 (53.1%)	83 (46.4%)	1 (0.5%)	179
Sibari	282 (53.5%)	222 (42.1%)	23 (4.4%)	527
Total	508 (49.8%)	435 (42.6%)	78 (7.6%)	1021

Table 20.7 Distribution of educational qualifications in the sample of visitors

Museum	No answer	Elementary school	High school	Baccalaureate	Master	Total
Altino	15 (10.9) %	15 (10.9%)	64 (46.4%)	10 (7.2%)	34 (24.6%)	138
Ferrara	15 (33.3%)	5 (11.1%)	11 (24.5%)	3 (6.6%)	11 (24.5%)	45
Ancona	32 (24.3%)	9 (6.8%)	43 (32.6%)	18 (13.6%)	30 (22.7%)	132
Egnazia	1 (0.6%)	16 (8.9%)	68 (38.0%)	24 (13.4%)	70 (39.1)	179
Sibari	31 (5.9%)	104 (19.7%)	148 (28.1%)	51 (9.7%)	193 (36.6%)	527
Total	94 (9.2%)	149 (14.6%)	334 (32.7%)	106 (10.4%)	338 (33.1%)	1021

1. Performance Indicators

From the data above, we can log the following performance measures:

1. The average approval rating of E.V.G. according to the visitors was 8.2 points out of 10 (scale of customer satisfaction)
2. The approval rating according to the Curators (assessed based on the availability of the same curators to recommend the "game" to other curators) was 8.5 points out of 10

Table 20.8 Distribution of professions in the sample of visitors

Museum	Professional	Business man	Artisan	Employer	Teacher	Student	Tourism expert	Manager	Other	No answer	Totale
Altino	8 (5.8%)	5 (3.7%)	2 (1.5%)	42 (30.3%)	14 (10.1%)	9 (6.5%)	0	5 (3.7%)	34 (24.6%)	19 (13.8%)	138
Ferrara	4 (8.9%)	0	1 (2.2%)	11 (24.5%)	2 (4.4%)	2 (4.4%)	2 (4.4%)	2 (4.4)	4 (8.9)	17 (37.9%)	45
Ancona	14 (10.7%)	1 (0.7) %	1 (0.7%)	33 (25.1%)	12 (9.1%)	16 (12.1%)	2 (1.5%)	2 (1.5%)	21 (15.9%)	30 (22.7%)	132
Egnazia	31 (17.3%)	0	9 (5.0%)	39 (21.7%)	26 (14.5%)	19 (10.6%)	1 (0.6%)	2 (1.2%)	46 (25.7%)	6 (3.4%)	179
Sibari	66 (12.5%)	7 (1.3%)	13 (2.5%)	112 (21.3%)	73 (13.9%)	123 (23.4%)	4 (0.7%)	9 (1.7%)	82 (15.5%)	38 (7.2%)	527
Total	123 (12.0%)	13 (1.3%)	26 (2.6%)	237 (23.2%)	127 (12.4%)	169 (16.6%)	9 (0.9%)	20 (1.9%)	187 (18.3%)	110 (10.8%)	1021

Table 20.9 The attendance of snouts in the sample of visitors

Museum	Less 5 (%)	Till 10 (%)	Over 10 (%)	No answer (%)	Total
Altino	75 (54.4)	37 (26.8)	13 (9.4)	13 (9.4)	138
Ferrara	16 (35.5)	13 (28.9)	4 (8.9)	12 (26.7)	45
Ancona	56 (42.4)	33 (25.0)	19 (14.4)	24 (18.2)	132
Egnazia	94 (52.5)	61 (34.1)	20 (11.2)	4 (2.2)	179
Sibari	298 (56.5)	127 (24.1)	74 (14.1)	28 (5.3)	527
Total	539 (52.8)	271 (26.5)	130 (12.8)	81 (7.9)	1021

Table 20.10 Frequency of use of the specific museum in the sample of visitors

Museum	Never	One time	More times	No answer	Total
Altino	91 (65.9%)	16 (11.6%)	19 (13.8%)	12 (8.7%)	138
Ferrara	8 (17.8%)	10 (22.2%)	14 (31.1)	13 (28.9%)	45
Ancona	70 (53.0)	24 (18.2%)	14 (10.6)	24 (18.2%)	132
Egnazia	139 (77.7%)	26 (14.5%)	11 (6.1%)	3 (1.7%)	179
Sibari	377 (71.6%)	74 (14.0%)	41 (7.8%)	35 (6.6%)	527
Total	685 (67.1%)	150 (14.7%)	99 (9.7%)	87 (8.5%)	1021

3. The approval rating of the curators regarding the relationship between the curators and the Dynamic Museums team was 10/10 points
4. 56.5% of respondents reported a preference for the logical characteristics
5. 43.5% reported a preference for the emotional characteristics
6. 33.4% of visitors indicated that they were 35 years and under
7. 58% indicated that they were 36 and over
8. 49.7% indicated that they were male, 42.6% that they were female and 7.7% declined to answer.

1. Examples of routes

The following figures show some examples of paths that have been carved out according to the needs expressed by individual visitors. Figure 20.1 depicts exhibits at the Museum of Altino, Fig. 20.2 pieces from Ancona Museum and Fig. 20.3 pieces from Egnazia Museum.

The following are the overall data (from the period of official testing and the subsequent period, that is, the "Canonic" period and what we have called the "Continuous" period).

Courtesy National Archaeological Museum of Altino

Fig. 20.1 Example of a proposed route, according to the needs of a given visitor to the Museum of Altino

Fig. 20.2 Example of a proposed route, according to the needs of a given visitor to the Museum of Ancona

Courtesy National Archaeological Museum of Marche

Courtesy National Archaeologics Museum of Egnazia

Fig. 20.3 Example of a proposed route, according to the needs of a given visitor to the Museum of Egnazia

ALTINO (02.08–22.09)

Logical characteristics	Emotional characteristics
12,393	8321

Age	Sex	School	Profession	Museums/year
Up to 18	Male	No answer	No answer	Less than 5
21	109	23	24	142
Up to 35	Female	Obligation	Freelance	Up to 10
47	119	24	12	70
Up to 55	No answer	High school diploma	Dealer	Over 10
91	20	119	7	17
Over 55		Graduation	Artisan	No answer
66		34	7	19
No answer		No answer	No answer	No answer
23		48	75	248
			Professor	
248	248	248	19	This museum
Age			Student	
Up to 18			20	Never
			Tourism	184

(continued)

(continued)

Age	Sex	School	Profession	Museums/year
			0	One time
			Entrepreneur	20
			8	Many times
			Other	22
			76	No answer
			Profession	22
			248	Museums/year
				248

Pieces and preferences

1	22_1_fornasotti_antefissa_potnia_theron	218
2	251476	39
3	251482_fornace	6
4	251598_stele_architetto	34
5	251600_stele_donato	4
6	al225	2
7	251410	5
8	251681	5
9	26_1_collana_d_oro	138
10	al6176_6178	7
11	al6291	43
12	al7262b	20
13	al7263	6
14	bassorilievo_nave_oneraria	10
15	capannone_del_latte_ollette_ovoidali	170
16	lastra_marmorea_dioniso_e_satiro	37

ANCONA (09.08–14.09)

Logical characteristics	Emotional characteristics
4357	3466

Age	Sex	School	Profession	Museums/year
Up to 18	Male	No answer	No answer	Less then 5
3	54	34	32	60
Up to 35	Female	obligation	Freelance	Up to 10
38	54	9	14	33
Up to 55	No answer	High school diploma	Dealer	Over 10
38	30	45	2	19

(continued)

(continued)

Age	Sex	School	Profession	Museums/year
Over 55		Graduation	Artisan	No answer
29		19	1	26
No answer		No answer	No answer	No answer
30		31	33	138
			Professor	
138	138	138	12	This museum
			Student	
			16	Never
			Tourism	71
			2	One time
			Entrepreneur	25
			3	Many times
			Other	16
			23	No answer
				26
			138	
				138

Pieces and preferences

1	anellone_a_sei_nodi_cupra_marittima	10
2	ansa_mobile_di_situla	28
3	ansa_hydra_picena	14
4	augusto_capite_velato	36
5	corona_montefortino_arcevia	14
6	cratere_attico_a_figure_rosse	73
7	deinos_di_prometeo	40
8	elmo fermo	26
9	fibula_a_navicella	11
10	leone ambra	2
11	Mosaico	7
12	orecchini_in_oro_cavalli_alati_celti_motefortino_d_arcevia	70
13	pettorale_a_pendaglio	39
14	rython_attico_a_figure_rosse_necropoli_sirolo_numana	31
15	sarcofago_di_medea	7
16	stele_funeraria_ancona	6

EGNAZIA (03.08–15.10)

Logical characteristics	Emotional characteristics
8618	6771

Age	Sex	School	Profession	Museums/year
Up to 18	Male	No answer	No answer	Less then 5
15	160	11	18	147
Up to 35	Female	Obligation	Freelance	Up to 10
60	151	21	45	123
Up to 55	No answer	High school diploma	Dealer	Over 10
119	9	112	0	37
Over 55		Graduation	Artisan	No answer
116		54	12	13
No answer		No answer	No answer	No answer
10		122	64	320
			Professor	
320	320	320	36	This museum
			Student	
			27	Never
			Tourism	256
			0	One time
			Entrepreneur	36
			7	Many times
			Other	15
			111	No answer
				13
			320	
				320

Pieces and preferences

1	1_bollitoio	8
2	10_oinochoe	29
3	11_ubriaco	17
4	12_banchetto	8
5	14_mg_0363	3
6	18_laminette	7
7	19_attis	243
8	194_sedia_mg_0605	19
9	2_macina	265
10	21_mosaico_grazie	74

(continued)

(continued)

11	23_fibula	75
12	27_salvadanaio	32
13	29_dadi	39
14	3_pugnale	141
15	5_guerriero	0
16	9361_dipinto	0

FERRARA (03.08–09.09)

Logical characteristics	Emotional characteristics
1418	1254

Age	Sex	School	Profession	Museums/year
Up to 18	Male	No answer	No answer	Less then 5
3	19	17	20	18
Up to 35	Female	Obligation	Freelance	Up to 10
11	17	5	4	16
Up to 55	No answer	High school diploma	Dealer	Over 10
13	15	13	0	4
Over 55		Graduation	Artisan	No answer
9		5	1	13
No answer		No answer	No answer	No answer
15		11	13	51
			Professor	
51	51	51	2	This museum
	Sex	School	Student	
		No answer	3	Never
		17	Tourism	10
			0	One time
			Entrepreneur	13
			2	Many times
			Other	14
			6	No answer
				14
			51	
				51

Pieces and preferences

1	1	1
2	10	3
3	13	6
4	14	5
5	9	3
6	17	10
7	18	14
8	20	1
9	21	6
10	22	13
11	24	21
12	25	13
13	26	17
14	32	12
15	4	25
16	8	3

SIBARI (03.08–27.10)

Logical characteristics	Emotional characteristics
7122	5557

Age	Sex	School	Profession	Museums/year
Up to 18	Male	No answer	No answer	Less than 5
124	396	54	61	420
Up to 35	Female	Obligation	Freelance	Up to 10
175	312	155	95	168
Up to 55	No answer	High school diploma	Dealer	Over 10
230	43	223	15	105
Over 55		Graduation	Artisan	No answer
178		78	22	58
No answer		No answer	No answer	No answer
44		241	150	751
			Professor	
751	751	751	91	This museum
			Student	
			166	Never
			Tourism	510

(continued)

(continued)

Age	Sex	School	Profession	Museums/year
			6	One time
			Entrepreneur	112
			13	Many times
			Other	72
			132	No answer
				56
			751	751

Pieces and preferences

1	MAS 03 PETTINE	92
2	MAS 06 ANFORA	79
3	MAS 07 DIPINTO DA VASO	391
4	MAS 08 MONETA	31
5	MAS 09 BASSORILIEVO	62
6	MAS 10 GIOIELLO	309
7	MAS 16 SCULTURA	115
8	MAS 17 INCISIONE SU TAVOLETTA	11
9	MAS 19 SCULTURA	147
10	MAS 21 ARMATURA	303
11	MAS 22 SCULTURA	57
12	MAS 23 ANFORA	39
13	MAS 24 VASO	145
14	MAS 27 VASO	40
15	MAS 29 UTENSILI	5
16	MAS 32 SCULTURA	427

Museum	Beginning date	Final date	Visitors	Evaluations	
	28.8.16	03/09/2016	With test	Comments	Score
Altino			138	119	114
Ancona			132	61	21
Egnazia			179	167	113
Ferrara			45	0	0
Sibari			527	464	108
			1021	**811**	**356**

20.6 Results from November 3rd, 2016

Risultati al 3 Novembre 2016

5 MUSEI TAVOLA FINALE   In fase di sviluppo

MUSEO	DATA INIZIALE	DATA FINALE	VISITATORI	VALUTAZIONI		voto medio numero	PRODOTTO
	2.8.16	03/09/2016	con giochino	con scritto	con numero		
ALTINO			248	119	114	8,19	934
ANCONA			138	61	21	8,90	187
EGNAZIA			320	167	149	7,17	1069
FERRARA			51	0	0	0	0
SIBARI			751	464	174	8,83	1537
						MEDIA TOTALE	
			1308	811	458	8,14	
							3727

Chapter 21
Contributions, Comments

21.1 New Participation in and Knowledge of the Museum

In recent years, the museum has been articulated as a protagonist of knowledge. There are a number of new strands that have opened up vast prospects for cultural and social development. The formal logic, artificial intelligence, and cognitive access to distributed information, and the huge amount of data that has been collected, including in the cloud, have changed the face of knowledge of the past, present, and even of daily events, expressed through cultural moments. The deeper culture has now acquired the technologies and methodologies that were introduced into science and human life in 2000. The main element is the widespread use of a distributed form of communication that spans the Earth's surface, with the knowledge distributed in bulk. Everyone is given the opportunity to participate in experiences from far away and share their own experiences with those just as far. Changes in the innovative spirit of the new world have resulted in a new form of human tissue. While computer knowledge has continued to produce useful tools for daily life and the application of virtual reality, and cognitive methodologists have invented formidable machines to enable culture to become pervasive everywhere throughout the world, a visit to a museum involving participation linked to moments of daily life required the creation of new paths. Science and technology have created mechanisms and systems that mimic the intellectual and emotional aspects of certain human functions. However, the sum power of human behavior, a combination of the logical aspects of the left hemisphere of the brain and the emotional aspects of the right, creates immense potential for billions of people to express themselves individually on this earth. Repeated and insistent efforts are made to pursue human behavior, which continues to fascinate us in its richness and mystery, through new mechanisms and electronic circuits. Today, we are interested in this new knowledge, which has been both exalted and debased by mobile phones, networks, databases, and the unstoppable power of your computer, which seeks to open new areas of knowledge and emotion starting, in part, from incomplete forms,

A. Rovetta and E. Rovida, *Scientific Knowledge Communication in Museums*, Research for Development, https://doi.org/10.1007/978-3-319-68330-0_21

towards the creation of something immensely and magnificently human. Thinking forward, the future may bring us a museum of knowledge that is increasingly super-globalized, forever producing richer, more complex souls, surrounded by exciting smart objects.

You will see, as we have seen so many times already, economic carnage and industrial crisis, causing changes in economic structures, a situation in which the cultural knowledge of the objects, as expressed by the museums, can have a positive effect, if the costs can be brought lower and lower. The logic of artificial intelligence is becoming increasingly conditioned through learning about human behavior, a fact that reinforces the indispensable necessity of the use of logic and understanding of emotional participation in human reality and the importance of investment in museum collections and events, particularly in their emotional dimensions, that will make these institutions more and more intuitive, understandable and beloved by people.

Alberto Rovetta

Presentation

The DYNAMIC MUSEUMS Project sprang from the observation that our Museums today, though often beautiful and visited by large numbers, suffer from a setting that, especially from the visitor's point of view, is a bit 'outdated.' The visitor, in other words, too often merely "walks" through the halls, observing, sometimes carelessly and with little involvement, the works on display.

The whole point of the DYNAMIC MUSEUMS project consists in putting the visitor at the center of the visit. This process can start either at a computer station at the entrance, or even at home, through use of simple electronic devices, such as computers, tablets, iPads, etc., with potential visitors expressing the characteristics, both logical and emotional, that describe their state of mind at the time. Cross-checking the pieces within the museum, which have themselves been assigned various characteristics by the curators, with the wishes of the visitor, a custom path is then "cut out" for the individual.

The project led to the creation of a platform that has been "tested" in five archaeological museums, selected by MiBACT, in Altino, Ferrara, Ancona, Egnatia and Sibari. The results were very positive: a total of 1600 visitors (and climbing) have "tried" the platform, and the feedback collected through special questionnaires has resulted in an 8.2 out of 10 grade.

The DYNAMIC MUSEUMS team, having noted the positive results with satisfaction, is preparing a development and expansion of the platform's applications, for use at a greater national level and then internationally.

Edoardo Rovida

1. This program is the answer for those who argue that it is time for the art world to discard the old rules, so that museums will be open to new ideas (The beauty of Italy—Corser).

2. The "Easy visit" platform, a form of innovative technological support based on neuroscience, can be of great use to the cultural heritage of our country, by placing the tourist at the center of their own visit.
3. The investigation by EY and Almaviva indicates that, by 2020, there will be a total willingness to utilize digital platforms within the context of tourism.
4. The tourist, having been given the opportunity to feel like the protagonist in their own story, one that is unique and unrepeatable, then turns into an "ambassador/propagator" for the program.
5. The program can also provide new motivational insights for the receiver that may become the focal point of relational development with the tourist.
6. The program can also be expanded to the surrounding territory, so that it all becomes a stage upon which tourists can live out memorable experiences as protagonists.
7. In these ways, the program becomes a tool for the innovative use of our cultural heritage as a generator of economic value and increased visibility for the country.
8. It is therefore suitable to celebrate a new "renaissance" in Italy, home to one of the original's greatest masters, Giovanni Bellini.

Giulio Zafferri—19/11/16

The DYNAMIC MUSEUMS Project comes from the need to fill a void: the lack of interaction between the visitor and the museum, and that between the visitor and the museum curator.

Most people are not aware of how many museums surround them, not to mention the many famous works that live inside of them.

Upon entering these places in which so many good works are stored, users (especially on their first visit) are out of place and are not always able to follow the paths indicated by the guides. In many cases, only at the end of the visit do the visitors discover, for example, by opening the catalog for sale in the bookshop, many of the works that would have attracted them, but which they failed to see for one reason or another, including lack of attention given to those objects by the museum itself.

DYNAMIC MUSEUMS focus on the emotions that human beings may wish to engage during their visit and how they might best request them and receive them.

People will have computer support through which to assess their state of mind according to numerical parameters. The mood will be divided between its emotional and logical characteristics, through a program that functions very much like the division that takes place during brain interface (and which produces an effect that has brought a smile to the faces of the users who have already tried the platform).

Previously, the curator would deliver logical and emotional values to each piece within the museum collection.

Crossing the data supplied by the User with that supplied by the curator will allow for the creation of a tourist path ad hoc.

In the five museums in which this platform has already been tried, the results have been phenomenal, with users spending more time in front of the sculptures or paintings that the platform suggested, developing a satisfactory emotional association with that material.

The project, currently still in its embryonic stage, is under constant development and will eventually be easily adaptable to various fields.

Vanessa Tritone

Courtesy National Archaeological Museum of Marche

21.2 "DYNAMIC MUSEUMS" Program: The Contribution to the Italian Economy Induced by the Dynamic Participation of Italian Museums

We believed that, using good and useful ideas, and the commitment of resources, an innovative project could be introduced that would not involve significant expense and would result in the enhancement of the resources of Italian art and its associated museums. Around the world, these resources are of unique and extraordinary cultural interest.

We submit a few simple considerations of a program that would open the magnificent Italian museum for greater exploitation of its resources, revealing new fronts for economic development.

We felt it would be good and useful to hear the opinions and gather the advice of those who dedicate themselves to the national cultural heritage and defend the enormous wealth that it represents, creating resources for the country, quantifying the number of Italian museums and monitoring their activities along the way.

Preserving a nation's artistic heritage not only keeps those resources useful for tourist activities alive and increases the number of entrance tickets torn, it also multiplies the revenue related to those sectors linked to it, such as transport, catering, accommodation, communication and merchandising. Many foreigners who might otherwise visit Italy by rote end up reaching places of great personal interest and creating indelible memories.

Unfortunately, far too much of Italian culture and its museums remains unknown by both foreigners and Italians alike.

The ambitious goal for which we aim is to increase income from tourism, make available a new type of visitor service that can be applied at various levels, and foster a museum culture that is sustainable and can benefit the surrounding area as well.

Basically, we wish to offer a New Cultural Service.

We decided to take a new road that would suggest to museums throughout Italy how they might improve their relationship with their visitors.

Engaging directly with more individual tourists, Italians, and foreigners, those who are cultured and interested in beauty, art and technique, we would suggest the objects and routes, the information and knowledge that could make their visit to the museum even more exciting. Digital technology is an integral part of this and would allow people to share information and feelings. PCs, Tablets and Smartphones, connected to the network, would considerably expand the catchment area of our application.

We have focused our energies on building a futuristic reality whose roots are the results of today and whose branches represent the results of tomorrow. The force of modern knowledge can overcome all obstacles of time, distance, environment and culture, if it proceeds through well-guided and precise objectives. We pulled out all the knowledge of computer design, communication design, data management, and virtual and multimedia representations of multivariate mathematics to show how more and more can be brought to each person within the museum reality. No longer will the set and programmed paths be the only ones considered; the free choice of knowledge and participation, fueled by intelligence and passion, will become a part of the museum experience as well. In this way, the visitors find themselves as the new front of the intelligent object, the work of art, the archaeological find, the scientific instrument. The quality of the items (as Aristotle wrote in 340 BC) drew human attention, in its logic, in its emotional and vital participation. It is a contribution parallel to that of the imposed and programmed path, which remains effective for a priori suggested knowledge.

21.3 Economic and Financial Effects of the Proposed "Dynamic Museums"

The DYNAMIC MUSEUMS proposal implicitly provides suggestions for employment development and use, to increase the operability of the Italian Museums.

The increase in jobs and related activities broadens the management of all museums, even the smallest ones whose reduced visibility means that they are very often known only in the immediate area. The most prominent Museums and sites can receive millions of visitors, as reported by documents out of the Ministry of

Heritage and Cultural Activities and Tourism (see attachments for 2015 and others), while hundreds of small state and non-state museums, which are also part of the Italian cultural base, struggle to attract visitors.

The DYNAMIC MUSEUMS Program would serve as a new model approach for Museums of all types, offering a most simple and intuitive tool to every person who wants to know everything that the nation's Museums must offer, a tool that will also be of great use to the museums themselves.

The new professionals needed will obviously include those graduates trained in the disciplines related to cultural heritage, but because this new and urgent work is all about reinforcing the links between museums and those citizens curious about culture, we will also need those well-versed in computer science, economics, management science, and tourism science.

Through the new methods of information technology and communication, museums can also achieve a valuable form of cultural updating, meaning graduates trained in the use of holography and 3D printers will also be necessary, to ensure the product is of high quality.

Graduates schooled in foreign languages will also be needed to promote the internationalization of the small Italian Museums.

Also, it is necessary to propose a method of integration in the use of the teachings of these new disciplines that will fit comfortably with all the assorted Museums.

The financial and economic aspects of the Project are linked to:

- Creating digital platforms to establish a tight Network between Museums, for now confined to those in Italy
- Creation of structures to make holograms for visitors who require them
- Facilities to make inexpensive 3D prints that will become fashionable
- Construction of tourist facilities related to the new Musealità, with inclusion of Museum products in tourism management
- Consequent adaptation of links between local cuisine and accommodations to objects in the Museums
- Creation of tourist gadgets related to the Museum and its promotion
- Internationalization of small museums and Network Management.

The economic impact can be evaluated only as a first approximation. Considering, according to the new program, increases of 2% for the flow of tourists and 10% high density areas (in small numbers) in the low-density area cultural Museum, the increase is attributable to the DYNAMIC MUSEUMS program.

Careful economic analysis identified, in 2015, an income related only to inputs from national museums, which amounted to 43 million euros, a figure that is equivalent to the turnover of a good average Italian firm. However, the resources related to tourism in Italy are closely linked to the activities of museums, and it is to be considered that the activities carried out by tourists are the most varied.

We have prepared a small evolution in the idea of the museum visit, which can, as sometimes happens, become pervasive and global. The small development involves the use of new knowledge about psychological neurology, something that has long been well-known within the scientific community, but that had gained little attention until a few years ago.

The "DYNAMIC MUSEUMS" Program starts from the knowledge of the activities of the two hemispheres of the brain, the left one for logic and knowledge, and the right one for the elaboration of perceptions and then feelings, emotions, and emotional involvement. For nearly five years, activities related to the hemisphere of emotions have been designed and developed with this new knowledge.

Should not the same apply to Museums? The reality of our great Italian riches? Why not turn new knowledge of our intelligent nature into a cost-effective approach to our national reality? Therefore, we have used all our knowledge about communication design, image management, the economics of information, the use of multimedia techniques and virtual reality, artificial intelligence, and multivariate mathematics, so far used mainly for the benefit of technological and scientific projects. We have focused our eyes (and our hearts) on the intelligent objects that are located, and often hidden, within our museums, both the large ones in big cities, and the small ones, even those in unknown villages, those of the sea and the mountains and the plains, with a method that is both "aggressive" and pleasant.

The method is very pleasant because it asks each visitor, as an individual person, to read a simple table depicting logical activities that a person may engage in at any time of life, along with the emotional activities, and corresponding perceptions, that also play such a large part of life. The tables allow the individual to register their interest in logical evaluation and emotional involvement with a score from 0 to 10. The simple, colorful table can be accessed on their mobile phone, and then transmitted to the DYNAMIC MUSEUMS program control computer. The table, for the sake of privacy, asks only age, sex, occupation, country, and, if the user wishes, an abbreviated nickname. The Visitor table is then distributed to be considered by all the museums in the network of the DYNAMIC MUSEUMS Program.

The Curator of each Museum has prepared a selection of significant objects, divided into groups of 8. Using a simple file, lightweight and fruibilissimo, it shows the image of each item proposed. The same table is compiled by the Museum Curator (or a resident expert/delegate), indicating the logical importance and emotional interest that the Trustee believes is transmitted through each intelligent object on display, also according to values between 0 and 10. This can be done in as many data packets as the Curator of the museum believes is necessary.

Once ready, the control center computer cross-references the Guest data with the information on each item that the Museum Curator has highlighted, using multivariate mathematics and the available neural network systems. In a matter of milliseconds, items from thousands of Museums are reviewed, and the visitor receives an ordered and precise classification of objects which would fulfill him and give him the pleasure of knowledge and emotion, selected from all the connected Museums. In a few seconds, the network provides the Visitor, who may be located thousands of kilometers away, a ranked list of favored items, which are selected and proposed for her and her alone. It serves as a technological gift to his personality, her intellect, his knowledge and her pleasure.

The beauty of the result emerges from a table, created by the network computer, indicating those Museum objects that correspond to the person's interests, and indicate the location of the Museum, the dates and hours at which it can be visited, the geographical position of nearby restaurants, available means of transport, and all the facilities required. The system provides an output package of 10 or 100 or more addresses, suggested according to the requests of each person. This transverse dimension of museum visits offers a much deeper use of the museum heritage, which is no longer simply consolidated for decades or centuries of acquaintances, but also offers the opportunity for new actors and creators to offer new products, for greater cultural richness.

The user does not need to move from their home or place of work, or even download the completed file. They also have the option of receiving all the information by mail, with a small payment for reimbursement of expenses. The user may then request a hard copy of the network catalog, with descriptions of the favorite items, and then decide which Museums would be preferential for the visit. In further indications for free choice, the system would serve the additional function of measuring the distance needed to travel, considering the operating hours, against the richness of the museum's offerings, and thereby the amount of time that may be needed to view its collection, to calculate the visit ergonomically and thereby make it more positive and productive, as well as economically viable. The visit that follows would constitute the "winning moment" of the "DYNAMIC MUSEUMS" program.

The main goal of the Project is to create a network joining all interested museums throughout Italy and making the corresponding database available through the Ministry of Heritage and Culture and Tourism. The Ministry can use the assessments in the software program to define the importance of each museum.

The database can then be used to affect a general search of Museums that takes no account of the historical importance of the museum or its collections, or for research that is influenced by decision parameters set by the network designer. Either can be implemented by each user, with different results, of course, depending on the user's choices.

One of the great positive effects of the "DYNAMIC MUSEUMS" program is the opportunity for the museum designer to modify and update the Museum's presentation based on the requests and responses of the visitors, who represent the true quality assessors of a museum. Thus, the Museum Designer has a perfect source of information culled from the analysis of visits and comments by visitors, even those that are merely online. We will now show an example of such an application.

21.4 Application Example

For each Museum chosen, the program will also include a link to available transport and tourism activities, with suggestions shaped by the parameters input by the user, who will then choose the most pleasing offerings. The objective of this application is to network all significant Italian Museums, interpreting and expressing their many exhibits (including those in storage) through usable data on the network submitted by as many users as possible, to which the visitor can refer before each Museum visit.

The first step in the "DYNAMIC MUSEUMS" program was constituted by the choice of 5 National Archaeological Museums, selected by MiBACT and Arch. Manuel GUIDO, with the collaboration of Dr. Valentino Nizzo, who examined cards related to all their objects, with evaluations of their logical and emotional aspects. The Support Center Computer then cross-referenced these data with the data of the potential visitor. A list was then released with a ranking of suggestions, which would allow the visitor, who has also given their own logistical and personal conditions, to make a rich choice. Each museum may make available more files with 8 objects each, since the design stress is very low and the processing time minimal.

This lightning-fast exploration is the result of the sort of technology that is easily accessible in 2017.

A short numerical and quantitative assessment of Italian Museums tells us that we have 202 state museums and 223 monuments and archaeological sites combined, which were collectively visited in 2015 by 43 million people, with approximately 155 million Euros having been collected. In total, there are about 4000 Italian Museums, supported by individual regions or provinces, some of which are private or belong to foundations. Some of them are beautiful and rich, their absence of visitors often being due solely to a lack of visibility and promotion, since many of them are not suitably "wired." This is not something that should be permitted to happen in 2017, particularly not for such cultural richness that is specifically Italian and is the potential source of wealth and income.

In fact, the economic value induced by Museums is tied to the cultural Visitor, with the Museum representing the endpoint of curiosity and pleasure, built up with all the other aspects of tourism and accommodation that bring money to the area. In 2015, foreign visitors to Italy spent billions of euros to meet their cultural tourism needs, an amount well beyond the purchase of one ticket at the Museum.

It is useful to compare the Program's system with the work being done by certain technological companies, and within the field of robotics. The products created within robotics, such as systems and infrastructure, now account for around 4 billion Euros per year in Italy. The value induced by the museum, in the minimum case scenario, is much greater, and taking in mind those who wish to experience the world of culture suggests a figure of 80 billion Euros, about 5% of the Italian GDP.

Our work regarding the Museum unit is only the starting point of a virtuous circle that could satisfy all visitors who wish to visit natural environments, cities, or even new countries, completing the experience of a cultural identity as part of a beautiful way of life. The DYNAMIC MUSEUMS Program asks the visitor about such preferences as visiting towns or cities, whether he would like to find museums in unknown locations or is indifferent to them, whether she gives priority to her intellectual or emotional curiosity. The program's calculation will then change the rankings of selected objects and make suggestions on the basis of these guidelines, providing a more enjoyable experience for the visitor.

The first consideration that arises is related to the advantage offered by the DYNAMIC MUSEUMS method. First, any human being who can read and write (and in the future, given the rise of simple icons, even without knowing how to read and write, people would still be able to participate) can fill in the tables regarding their logical and emotional interest on a mobile, tablet or computer, as well as providing data such as sex, country, occupation, and main fields of museum interest, if desired. After a few seconds, the DYNAMIC MUSEUMS system suggests 10% of museums that match the liking of the Visitor. The visitor has now joined the number of possible users of those museums indicated, and if the presentation is appealing and attractive, the virtual and remote user can decide to visit the museum and enter the museum object and museum into his tour route. Of course, the system also provides travel distance and possible points for accommodation, food and support for the visit in the chosen venue. Thus is born a welcoming path, which is user-created, turning the user into a Guest. It is quite easy to assess the economic value induced by this virtuous circuit, both for the venues who will receive her and for the Visitor, who, having built her own personal journey, will feel the satisfaction of getting exactly what she has requested. In addition, the restaurant business and local transport are mobilizing the resources that make the Museum and its riches more utilizable as well.

With the "DYNAMIC MUSEUMS Program," the user can see all the suggested items and choose them, or settle for displays on their mobile screens and computers, reading all related information. The Museum's inherent interest provides the best documentation possible to acquire a new customer. The object can also be presented with holograms, models made with a 3D printer, video, files, or multimedia communications, at the lowest possible cost.

In conclusion, even limiting ourselves to potential Italian customers, a virtuous cycle could be triggered within at least 10 national museums. All that would be necessary would be the development of a simple program for the Visitor and one or two cards for each data management Museum, prepared by the Curator of the Museum. The final balance sheet is then prepared, along with a ranking of the selected objects with a figure and an accurate description of the object. Once the potential visitor has chosen their favorite objects, the program suggests the best paths for them to visit the museums, including those that are most economical and effective, the local hotels and facilities suggested by other users, according to its data. Moreover, the visitor will receive holograms, virtual objects or models built with a 3D printer, based on the available technological properties.

To accomplish this in the first "season," you will need a team of 5 people, constituting the basis for a start-up, with 2 communication designers, 2 computers, and a manager. In one year, we have three phases: (1) Preparation of complete and flawless software (2) Delivery of an app with a selected sample of users (3) Tests and results with early indications on the success and improvement of the visit. Furthermore, funds must be available for the Curators of the project and for the dissemination of the results through the media and the press.

The funds can be 150 K Euro for start-up for a year, with 5 people, and 100 K Euro for the management of results and design of the main programs. A first call for funds will be sent out, with a strategy of small steps and immediate verification of results, to foundations and banks, including to European projects of immediate access. In addition, it is necessary to establish preferential contact with selected departing Museums and along a special line, one that will always remain active, which stores all data and results, for the construction of an interactive and dynamic design. A book will also be provided for the international dissemination of the DYNAMIC MUSEUMS Project, and is already being prepared by Springer Verlag, a world leader in scientific and technical publishing. Control of the project is entrusted to a team of experts, restricted to 7 persons, composed of directors of Italian Museums and designers of international fame.

In conclusion, the objective of the DYNAMIC MUSEUMS project is to offer a service to the Italian culture, with an amplification of the management of Italian Museums, including those that are less well-known, for the maximum exploitation of the cultural heritage. The program opens new horizons to non-expert museality, and therefore becomes a lever for increasing tourism, through the redefinition of cultural and museum tourism. Also, it will certainly result in customer satisfaction, i.e., the approval of the Visitor, for they are only offered those objects that matter most to their cultural and emotional perception. It allows them to obtain information on their favorite museum objects without cost, except for the small fee for participation in the App. Moreover, the Museum Curator, based on the characteristics of the new Visitor, can evaluate any potential redesigns of the museum, as is already happening in the Museums participating in the pre-analysis.

This new cultural service has shown that there is a new way to manage the fruition of museums, which are a rich mine, as well as a breeding ground for the well-being of the Italian culture, and a benefit to visitors and tourists from around the world who are in the network App of DYNAMIC MUSEUMS and who want to make their way to Italy. The proposed model is simple, and opens the door, one that is sometimes not always available, to thousands of beautiful small and large Italian museums, which are golden objects of the Italian

Peninsula and should be donated as symbols of luck and economic destiny. The proposal is dedicated to creating a new cultural service for our Museums in Italy, a new methodology to provide more extensive visibility to contemporary objects and places of culture, an intuitive sharing of the pleasure to know and experience beauty, the possibility of using services for ever better and effective cultural tourism. Simply put, we can adapt ourselves to the new reality of cognitive participation at very low cost, when compared with the growth of the use of Italian museums, with the added advantage of economic development of the areas where the museums are located.

It also offers cultural growth, combined with those mathematical models for digital communication that pervade our quality of life, in every moment, leaving most of us, especially those of us who have the desire to increase our knowledge, more satisfied and reaching increasingly towards the future. That future has a symbolic event taking place in 2020, the Expo in Dubai, where the motto is "Connecting Minds, Creating the Future," and the sub-themes are Sustainability, Mobility and Opportunity.

When the project and its results reach the critical threshold of successful participation, it will significantly add to this minimal contribution to the Italian presence in Dubai, a bridge between Italy Post Expo 2015 and the educated and bright future that awaits us. The Italian museums can represent the path of human civilization, with objects, signs and symbols that accompany us for centuries and guide us into the future (Figs. 21.1, 21.2 and 21.3).

Fig. 21.1 Map of National Museums with the figure from the Ministry's website

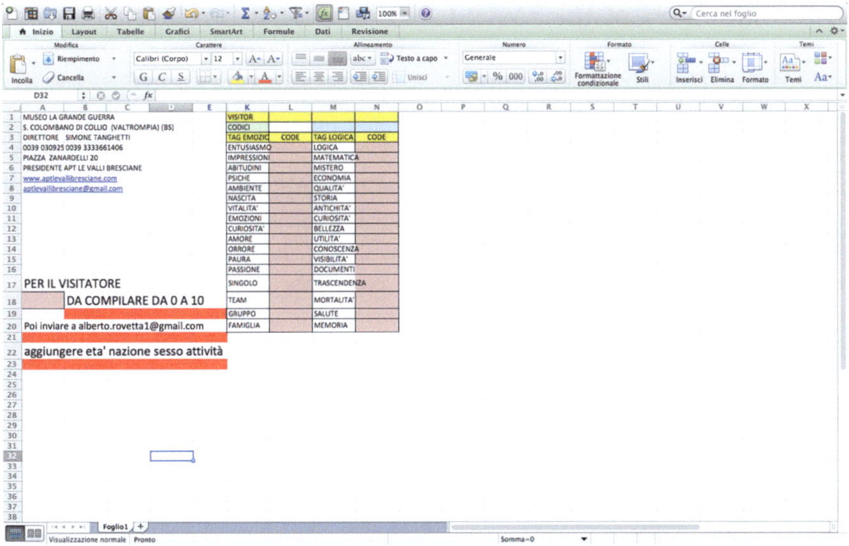

Tavola per compilazione da parte del Visitatore

Fig. 21.2 Table for phone, tablet or PC to be filled in by the Visitor, in seconds, with the support of basic instructions

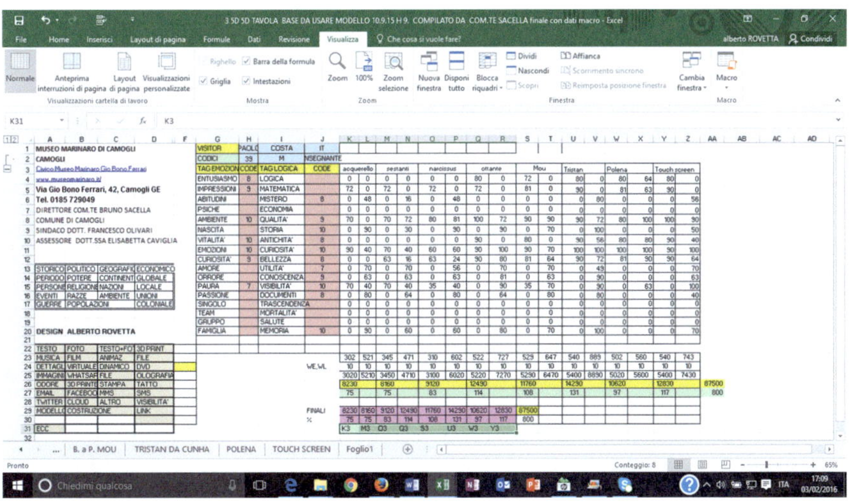

Tabella finale di sintesi

Fig. 21.3 Results from the cross-referencing program, which produces standings with a list of suggested items, according to the intersection of the mathematical data provided by the user and the data parameters assigned by the Museum Curator

21.5 Summary of the "MUSEI_DINAMICI" Project

The DYNAMIC MUSEUMS Project radically transforms the use of museums by involving the visitor according to his personality through logical cognitive aspects and through her emotions, expressed in daily life by the actions, respectively, of the left and right hemispheres of the brain. The MUSEI_DINAMICI project would enhance the convergence of logical and emotional interests of the Visitor with the participation, by electronic means, of the curator of the museum. You can use a computer, cell phone, tablet, or any other communication device, as well as a computer station in the Museum itself. Through mathematical algorithms, the MUSEI_DINAMICI Project generates a numerical-physical platform of suggested results for the Visitor, which lists the most pleasing objects of interest to the visitor in all participating museums. The platform also serves for the reconfiguration of the museum routes, which can be prepared for each Guest, or even the Museums themselves. THE NEW SERVICE OFFERED for the new use of the Museums takes place through identification of paths between multiple museums, offering all necessary information online for reaching and accessing these museums, along with all documentation of the local facilities that will make the visit more comfortable and appealing. The result for each Guest can also produce souvenirs in the form of printed photos, holograms, or 3D-printed models, which will better allow them to remember the experience they had with their favorite objects.

MIX
Papier aus verantwortungsvollen Quellen
Paper from responsible sources
FSC® C105338

If you have any concerns about our products,
you can contact us on
ProductSafety@springernature.com

In case Publisher is established outside the EU,
the EU authorized representative is:
Springer Nature Customer Service Center GmbH
Europaplatz 3, 69115 Heidelberg, Germany

Printed by Libri Plureos GmbH
in Hamburg, Germany